Environmental Division 2013

Core Programming Topic at the 2013 AIChE Spring Meeting &
9th Global Congress on Process Safety

San Antonio, Texas, USA
28 April – 2 May 2013

ISBN: 978-1-62748-460-2

Printed from e-media with permission by:

Curran Associates, Inc.
57 Morehouse Lane
Red Hook, NY 12571

Printed by Curran Associates, Inc. (2013)

AIChE
3 Park Avenue
New York, NY 10016-5991

Phone: (203) 702-7660
Fax: (203) 775-5177

www.aiche.org

Additional copies of this publication are available from:

Curran Associates, Inc.
57 Morehouse Lane
Red Hook, NY 12571 USA
Phone: 845-758-0400
Fax: 845-758-2634
Email: curran@proceedings.com
Web: www.proceedings.com

TABLE OF CONTENTS

(27a) Life Cycle Assessment and Life Cycle Cost of the Pyrolysis of Switchgrass to Use As an Energy Source in a Conventional Power Plant

Nuttapol Lerkkasemsan, Chemical Engineering, Virginia Polytechnic Institute and State University, Blacksburg, VA

Biomass is expected to become an important energy source. This paper considers both life cycle assessment and life cycle cost of the pyrolysis of switchgrass to use as an energy source in a conventional power plant. Specifically the greenhouse gas (GHG) emission for power generation was investigated through life cycle assessment. The process consisted of cultivation, harvesting, transportation, storage, pyrolysis, transportation and power generation. Here pyrolysis oil is converted to electric power through co-combustion in conventional fossil fuel power plants. The conventional power plants which are considered in this work are diesel engine power plant, natural gas turbine power plant, coal-fired steam-cycle power plant and oil-fired steam-cycle power plant. Several scenarios were conducted to determine the effect of selected design variables on the production of pyrolysis oil and type of conventional power plants. The set of design variables consisted of land fraction, land shape, the distance needed to transport switchgrass to the power plant, the distance needed to transport pyrolysis oil to where, and the pyrolysis plant capacity.

The results show that with an existing average agriculture land fraction in the United State which is 44.1 percent of all the United State land area, the cost of electricity from pyrolysis of switchgrass is lowest at $1.2 per gallon with a capacity of 5000 ton of switchgrass per day of pyrolysis plant. However, power generation from switchgrass required huge land especially for an existing average agriculture land fraction which is 0.441. In order to provide enough switchgrass for a pyrolysis plant capacity of 5000 ton per day, a large land area with a size of 1706.11 square miles is required; note that the land area is bigger than the state of Rhode Island. Using natural gas turbine power plant for electricity generation, the price of electricity can go as low as 7.70 cent/kwh; contrast this with the average electric price of 10 cent/kwhin the United States. Based on 5000 tons of switchgrass per day, a natural gas turbine power plant can produce 5011.86 million kilowatthours, which is enough to supply the state of Hawaii. The main advantage in using a pyrolysis plant is the negative GHG emission from the process. This conclusion came from Life cycle assessment analysis of power generation using pyrolysis oil, which showed that it was environmentally friendly and would therefore reduce the GHG emission to the atmosphere. If and when carbon credit becomes significant, pyrolysis oil could be even more competitive.

(27b) Natural Gas as a Transportation Energy Resource: Life Cycle Considerations for Personal Mobility Options

Christian M. Lastoskie, Civil & Environmental Engineering, University of Michigan, Ann Arbor, MI

Presently the U.S. is experiencing a significant boom in domestic natural gas production due to the application of hydraulic fracturing and horizontal drilling to Marcellus and other formations with an abundance of shale gas. The combination of decade-low natural gas prices and increased restrictions on coal utilization for electric power has led to a displacement of coal-fired generation with electricity from natural gas.

With the prospect of the United States undergoing a transition to being a net energy exporter within the next decade, it is timely to consider scenarios wherein a displacement of petroleum as the predominant domestic transportation energy resource occurs, with steadily increasing utilization of cheap abundant natural gas as the source fuel for personal mobility. In this presentation, we consider three options for natural gas-derived personal mobility: compressed natural gas automobiles, battery electric vehicles, and hydrogen fuel cell cars. The life cycle environmental impacts are compared for the operation of these three types of vehicles, including the fuel processing infrastructure required to deliver the respective fuel for each vehicle. We consider the impact of potential future constraints on carbon emissions on the environmental footprint of each mode of natural gas-powered personal mobility, on a per mile driven basis.

(27c) Environmental Controls in Coal-Based Electrical Energy Production

Noah Meeks, Research & Environmental Affairs, Southern Company Services, Inc., Birmingham, AL

Over the past half-century, the coal-fired electric power plant has evolved from a simple machine to a complex chemical process plant. While the heart of the industry remains the same: turning turbines with coal-generated steam, the treatment of coal combustion emissions has grown considerably and more processes are under development. There is a significant role for chemical engineers to play in meeting these current and future needs. This tutorial will cover the environmental impacts and pollution control unit operations of coal-fired electric power plants, both current and future. The four main unit ops are selective catalytic reduction of NOx, electrostatic precipitation, fabric filtration ("baghouse"), and flue gas desulfurization ("scrubber"). These incorporate a number of chemical engineering principles including separation processes and reactor design. In addition, new compliance drivers are pushing additional pollution controls for mercury, which is typically done as an adsorption process. Finally, concerns over global climate change have resulted in extensive research into carbon dioxide capture and sequestration using absorption processes. Also water and wastewater impacts from the power plants will be covered.

From Research to Practice: A Story of Mathematically Predictive Actionable Process Safety Leading Indicator Index

AICHE - 2013
Abstract ID: 294440
Submitted to: ENVIRONMENTAL DIVISION

Author Contact Details

Speaker: Joe Stough, **VP Innovation & Technology**
Company: IHS
Email: joseph.stough@ihs.com
Phone: +1 713 499 4288

Author Bio:
Joe Stough is the visionary and strategist behind the IHS Operational Excellence Forum methodology. With a professional background in large-scale database software design a graduate degree in Applied Statistics from the University California, Stough founded Syntex Management Systems and led the company in providing solutions that successfully combine the disciplines of enterprise-level
business software design and advanced statistical analysis to dramatically impact Enterprise Risk on a global basis. Syntex is now a part of IHS, Inc., where on an annual basis, Stough is continuing to refine his findings in the context of the Operational Excellence Forum.

ABSTRACT:

For the past several years, a group of global Energy companies, including nearly all of the super majors, has been working on a research initiative with a common mission to find an actionable leading indicator index to drive Safety and eventually overall Operational Excellence performance.

This group has been collaborating to provide a large multi-year data set, which contains millions of data records from events such as incidents, investigations, near misses, audits, observations, assessments, and many other routine field-level activities. Since 2008, a rigorous statistical analysis process has been iteratively applied to this data to identify the mix of leading metrics which most effectively predict Safety performance outcomes.

This paper will review the leading indicator research findings drawn from analyzing this large Energy industry data set. It will reveal the key components of an actionable leading index which has been found to be uniquely strong in organizations who continue to sustain top-level Safety lagging performance.

From the analytical research and careful review with this consortium of industry QHSE executives and Subject Matter Experts, the index that was found to best predict the Safety outcome performance of an organization included measurements of (a) proactive event reporting (Reporting Culture), (b) the discipline and consistency in execution of QHSE business processes, (c) the timeliness in closure of important corrective / preventive actions, and finally (d) the responsiveness of supervisors and line managers.

The paper will provide an overview of the analytical process that produced these results and will outline alternatives for applying these insights on leadership performance dashboards. Although the leading indicator research and analytics activities continue annually to strengthen the effectiveness of the leading indices, companies are applying the index to provide actionable leading metrics on executive scorecards. We will outline a sample process for implementing practical, "actionable" leading metrics in each key area to drive line manager / leadership behaviour and ultimately improve QHSE performance outcomes.

Overview and Background: Researching and Solving the Problem

An organization attaining true Operational Excellence has combined best-of-class business and financial performance with zero loss of production up-time, asset integrity, the environment, and safety. However, the frequency and severity of operational losses are significantly above zero for most corporations. Contributing to this is the lack of a standardized measurement of operational loss to enable performance comparisons and benchmarking.

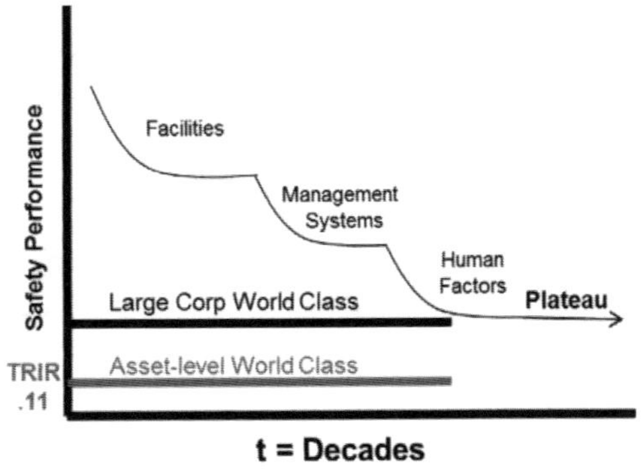

Although it is difficult to standardize the measurements of production up-time, asset integrity, and environmental outcomes across varying types of assets, the one area that does enable a relatively standard view of performance is **safety**. And since many industry executives hold the belief that the safest operating assets are also the best run operations, our research views the Operational Excellence problem through the lens of safety performance.

Figure 1: The Problem
Over the past few decades, the industry as a whole has seen safety incident frequency rates trend toward zero. But two key problems remain:

1. Even for the world-class performers, this trend has reached a plateau; and

2. The rate of high-severity injuries and fatalities has NOT followed the trend toward zero.

With these issues in mind, the focus is not only to drive Total Recordable Injury Rates (TRIR) to zero but more importantly to drive **severity-Weighted** Total Recordable Injury Rates (WTRIR) to zero. Whereas this goal of zero is perceived by most to be lofty at the aggregate or corporate-level, our research data set revealed operating assets (as depicted in Figure 1) which have sustained performance very close to zero.

In addition, the research data set revealed that the top 10% of assets as measured by WTRIR had the following interesting characteristics in comparison to the bottom 10%:

- WTRIR was roughly two orders of magnitude better
- Rate of near misses reported was 78% higher
- Incidents were reported 3 times more promptly
- Action items were authorized by leaders over twice as fast
- Rate of completed action items was 79% higher

Since these assets are sustaining near zero losses in the same risky environment with a similar level of safety exposures, this indicates the possibility of a much higher degree of Operational Excellence than is routinely achieved in industry.

A Path to Solving the Problem

With the top-performers raising the bar, a key challenge is to discover what the leaders, the workforce, and the overall organizations of these assets are doing differently. And the answer needs to come in a form that is measurable and actionable so that it can be repeated and applied to lower performing assets to continuously drive toward zero loss.

Specifically, the question that needs to be answered is: "What are the actionable, measurable factors which yield the near zero operational loss performance that is being achieved by these top performers?"

Using a severity-weighted safety incident frequency rate as the standardized outcome measure of operational losses, the research studies assets with performance varying from near zero to orders of magnitude worse to understand what creates high performing assets. The inputs in this research include data from workforce involvement in proactive activities such as audits, inspections, management system assessments, and observations as well as reactive events such as incidents, near misses, investigations and corrective actions.

By applying a rigorous statistical analysis process, the research reveals correlations between multiple leading factors and severity-weighted safety outcome performance. The focus is to discover those leading metrics which are not only predictive of outcome performance, but also are the most practical for use by top leadership as actionable, routinely measurable management controls. The process is continuously repeated with the aim of creating the optimum set of leading metrics which can be combined into a single, predictive multi-variable leading index.

The intended use of this index is two-fold:

1. To provide operating executives with actionable management controls for stewarding the process of continuously improving Operational Excellence performance; and
2. To deliver benchmarking of statistically proven leading indicators which reveal performance gaps and enable measurable goals and annual performance improvement plans.

Additional Obstacle: Management Addiction to Lagging Indicators

A primary obstacle to realizing the benefits of the above approach is top-level management's historical addiction to lagging indicators. Key Performance Indicators (KPI's) are management metrics used by business leaders to measure operational performance and to compel leadership to make performance improving operational changes. Lagging KPI's for QHSE are calculated from the occurrences of medium to high consequence loss incidents and provide a historical view of outcome performance. An often-referenced metaphor infers that using lagging indicators to drive performance is like "driving by looking in the rear view mirror". But in our research, it is extremely rare at the top-executive level to see anything other than lagging metrics of Safety performance used to judge performance.

Figure 2: Lagging KPI's from loss events

Per this depiction of the "Heinrich triangle", reporting data from incident events of the highest severity is routinely governed by a corporate standard (and often regulated) - thus rendering lagging data widely applicable and measurable across the enterprise.

In most companies, managers still use TRIR or a similar lagging metric as the primary KPI for judging an operation's Safety performance. One reason for this dominance is the practicality of having a standard in producing a normalized performance metric which can deliver an "apples to apples" comparison of these lagging KPI's across the enterprise (as described above). In addition to being routinely measurable, lagging measures are believed by management to be more quantifiable, tangible, objective, and generally more reliable than any leading KPI's proposed by the business.

Criteria of Leading KPI's as Management Controls

For business leaders to overcome this addiction and enable a leading KPI to truly rival lagging

KPI's as the preferred metric on a corporate-wide scale, the leading KPI must meet the following criteria to become a management control:

Actionable

To drive leadership behaviour, the leading KPI must offer the operational leader with a practical solution which can be acted upon to affect operational changes.

Objective

Surveys and opinion based assessments are too subjective. The best leading KPI's involve objective calcutions drawn from detailed activity or measurable events.

Normalized

TRIR enables and "apples to apples" comparison. Similarly, a leading KPI must be normalized to enable management comparisons across operating units.

Routinely measurable

The most effective and accurate leading KPI's are calculated from measuring leadership and worker involvement in routine operational activity on a relatively frequent (i.e. monthly) basis – as opposed to annual opinion-based surveys.

Believable and predictive

To attain the full commitment of field-level leaders, the leaders must believe that outcome performance can be affected with proper changes to performance in the leading KPI. Data and math can be used to assist leaders in developing this belief.

Throughout the aforementioned research and statistical process, the above criteria were applied to assure the resulting index of leading KPI's would likely be a cure. The rest of this paper will describe the research and business practices designed to overcome management's lagging indicator addiction.

Research Hypothesis

Per the premise described above, businesses are involved in the research to find the keys to preventing operational losses. These companies believe that an operational loss is a function of an organization's inability, either via failures in systems or behaviors, to effectively mitigate the risks involved in the loss. Therefore, the following is the research hypothesis:

"Operating units that are the best at reducing risks will sustain the best lagging Safety performance."

Risk Reduction Cycle (RRC) Database

An important characteristic of the research initiative is the availability of data pertaining to an organization's efforts, business processes to identify and **reduce the risks** to operational losses. The companies involved in the research have implemented a common enterprise-wide database application to consolidate data from business practices that fit the "risk reduction cycle" pattern. These companies are able to draw data from the database to observe the execution of these routine operational activities and are minimizing the effort of gathering leading data along the way. The following risk reduction process pattern of work practice activities applies to many QHSE business processes found in routine operations.

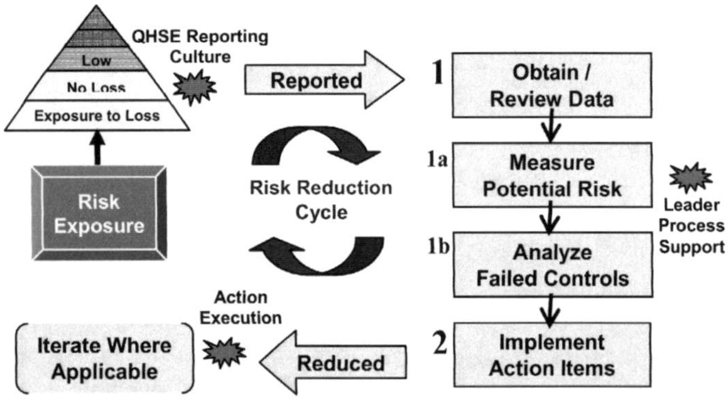

Figure 3: Risk Reduction Cycle pattern

As depicted, there are 2 key elements which are common to all risk reduction activities: (1) obtaining awareness via "reported" events and (2) implementing corrections to reduce the risk – i.e. "reporting and fixing" reported items.

Through researching operating facilities in the Energy industry, hundreds of event types have been discovered as risk reduction events – and thus contribute to the data set. The list ranges from multiple different types of incidents, near misses, and investigations to the many different types of inspections, observations, audits, etc...

The Core Elements of the Risk Reduction Cycle

The following describes how each element of the risk reduction cycle generally maps to the key steps in work practices resulting from both reactive (incident-based) events and proactive (assessment-based) events.

1. Obtain / review data (event reporting)
If you don't know about the risk, you can't reduce it. This element maps to the initial incident and assessment-based process steps whereby events are reported and people are assigned to contribute expertise / input and to review / approve the report.

1a. Measure potential risk (risk score)
For incident events, this is an optional / advanced practice of classifying "how bad it could have been" by using a risk matrix. Risk assessments, PHA's, and various other proactive processes include this as a key process step where the risk level is formally scored.

1b. Identify failed controls (root cause analysis / investigations)
For incident events, this element maps to the investigation process. Since many companies only investigate high-severity or high-risk events, this step is optional as well. An advanced practice is to apply an informal investigation to classify root causes and map those causes to Management System elements for ALL incident events – including near misses. For assessment-based processes, this element often represents the core purpose of the event – i.e. to identify the areas where risk control needs improvement.

2. Implement / repair controls (action item management)
For incident events and proactive events alike, this element maps to the most important step in all risk reduction processes – the execution of tasks to repair broken controls or implement new ones to ultimately reduce exposure to risk.

The research involves defining and calculating metrics across each of the above aspects of risk reduction process execution. Per the hypothesis, the intent is to identify the appropriate mix of metrics which both effectively measure an organization's risk reduction capability and are predictive of safety performance.

Researching a World-wide Data Set of Risk Reduction Processes

The companies involved in the research project provided their data on a scale that spans thousands of operating facilities in over 100 countries. Engaged in this initiative are world-wide operations from multiple Energy companies which have world-class QHSE performance... as well as leading companies in Oil Field Services, Chemicals, Manufacturing, and other industries where managing operational risks is a material concern.

Although most of the companies referenced in this study deploy unique company-specific programs for managing QHSE performance, they share one thing in common. Each of them is applying a common technology solution to collect and analyze data from a myriad of risk reduction process events such as incident investigations, near miss reports, management system audits, risk assessments, assurance reviews, behavioral observations, field-level inspection programs, hazard analysis, and many other processes. As depicted in the below diagram, the intended scope of this database is to be the organization's collection of all events that result in risk reduction actions.

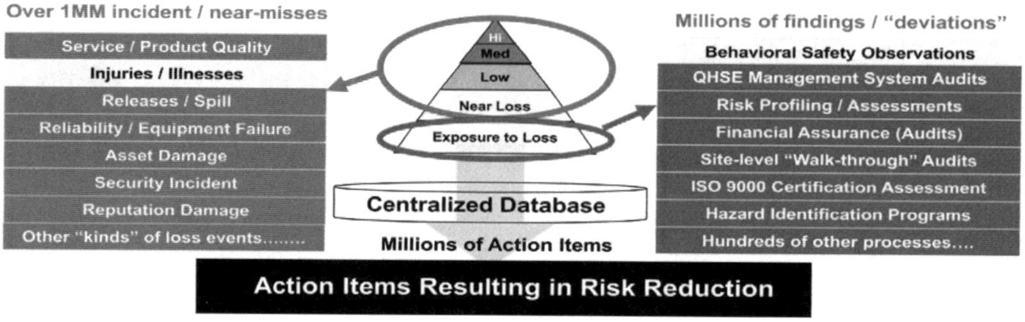

Figure 4: Risk reduction cycle events and actions database

On the surface, these organizations have simply been using a common "mechanism" to manage their own unique set of QHSE risk reduction processes and ultimately analyze the resulting data. However, at a deeper level, these companies are not only collecting data resulting from the event occurrences (e.g. incident reports, injury details, spill quantities, near miss types, root causes, audit results, assessment scores, inspection findings, etc.) but also the "work practice behaviors" reflecting the organization's tendencies in executing the processes behind such events (e.g. mean-times between completion of "critical" process steps, rate of leadership involvement, ratio of near-misses reported to high-consequence incidents, etc.). Applying statistical analysis to this vast data set has revealed the measurable organizational tendencies which are unique to top performers. The ultimate goal is to convince top business leaders to use these types of leading measurements.

From Research to Practice

Going forward in this paper, we will unveil some of the math-grounded process-based metrics which were discovered via the research initiative. Then, we review the critical role that each of these measurable areas play in the mission to proactively drive Safety performance. We'll end by showing the best approaches for applying these leading indicators on Key Performance Indicator (KPI) dashboards through an iterative measurement process to shift leadership focus, continuously improve QHSE performance, and attain Operational Excellence.

The Research Initiative: A Statistically Significant Data Set

Over the last several years, a collective data set of millions of records of data covering over a million direct and contractor employees of customer companies has been accumulated. As conveyed in the below diagram, the risk reduction cycle database manages both lagging and leading data. This has afforded the opportunity to study the relationships between lagging and leading data elements – with the purpose of finding statistically proven leading indicators.

The Opportunity to Correlate

The data set includes lagging metrics derived from high-consequence incidents as well as multiple sources of potential leading metrics including (a) leading events such as near-misses, low-consequence incidents, assessments, observations, action items, etc. and (b) the organizational tendencies in managing the business processes associated to such events.

NOTE: Although as depicted in Figure 4 low-consequence incidents and near misses are part of the incident event database, the organizational behaviours associated to reporting and managing such "non-mandatory" events are categorized in Figure 5 as leading areas of measurement.

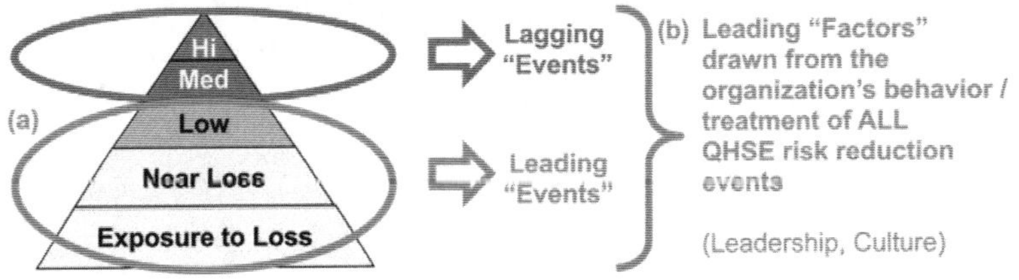

Figure 5: Sources of Data

With lagging and leading measurements coexisting in the same centralized database, a unique opportunity exists to practically, rigorously study correlations and identify math-proven leading indicators.

(a) A Database of Risk Reduction Events and Actions

As previously described, a risk reduction event is any event that starts with the (lagging or leading) discovery of a risk exposure and ends with the assignment and completion of action items to strengthen protections and ultimately reduce risk exposure. Examples of "lagging" events contained in the data set include near misses, incidents, and associated causal investigations from safety, security, environmental, quality, reliability (maintenance), and other types of loss events. Also included are "leading" events such as audits, inspections, behavioural observations, management system assessments, and over 400 other types of assessment-based events. Collectively millions of incidents, assessments, and action items have been accumulated over the past several years – establishing a rich, statistically significant data set for the research initiative.

(b) Data to Measure "how organizations manage risk reduction events"

The goal of these business process software tools is not only to collect the detailed data from risk reduction events but also more importantly to manage the full process lifecycle from initiation (reporting) of events through each key step in processing the event all the way to completion of the final corrective action. The research applies statistical analysis to this vast set of data to study the organizational tendencies in how the various parts of the organization are managing their roles in the risk reduction process.

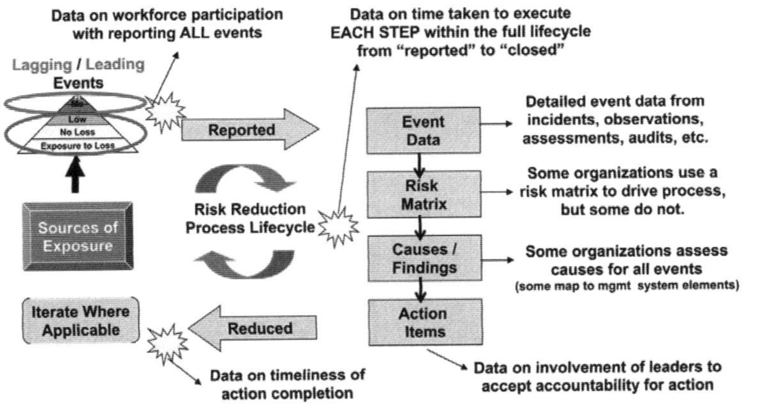

Figure 6: The Process-based Data

The diagram to the right depicts the many organizational touch points throughout the full "risk reduction process lifecycle" from which the metrics are being drawn. Through collaboration with Subject Matter Experts (SME's) from participating companies, over 200 different metrics have been defined in the research thus far – ranging from measures of reporting culture to various types of leadership and process-based metrics to measures of action item performance.

Research Premise: Find Opportunities for Improvement

The premise of the research initiative is to study relationships between the ordinary lagging metrics and the multitude of leading metrics as described above. The objective is to reveal measurable differences in the organizational factors which convey how organizations treat risk reduction events. These differences are reflected by substantial variation in metric scores when calculated across a group of organizations. Although the vast data set (described above) includes many different types of data elements for study, it is the leading metrics which expose such differences that represent the truly meaningful opportunities for improvement.

Getting to the Meaningful Data

The result of an enterprise-wide implementation of such a risk reduction process database is a practical means of collecting and analyzing data spawned from the event processes within the scope of the project. After establishing such a platform, some company executives have asked the following question:

"How do we get the <u>meaningful data</u> to enable our managers to lead performance improvement?"

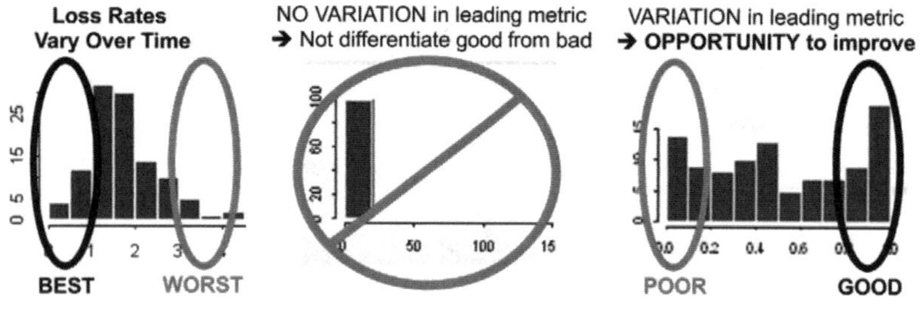

Figure 7: Meaningful Data

The best leading metrics are the ones for which the top performers (in terms of loss rates) are doing well and the worst performers are doing poorly. Because this variation exists in loss performance, a meaningful leading metric must show variation (e.g. example metric to far right) to represent an opportunity to improve.

Answering the Question: "What are the BEST performers doing differently?"

As conveyed in the above diagram, the histogram to the far right shows a metric with substantial variation – i.e. a material number of organizations have scores in each major segment of the performance spectrum. In this example metric (near misses as % of total incident events), organizations with a score near zero show up on the left side of the histogram and those with a score near 100 show up on the right side – i.e. some good performers and some bad.

For the best leading metrics, the majority of those organizations which show up on the "good" side of the metric's histogram are also present on the list of the BEST performers in terms of loss rates. These metrics represent the unique measurable characteristics of top performers and are the target of the research initiative.

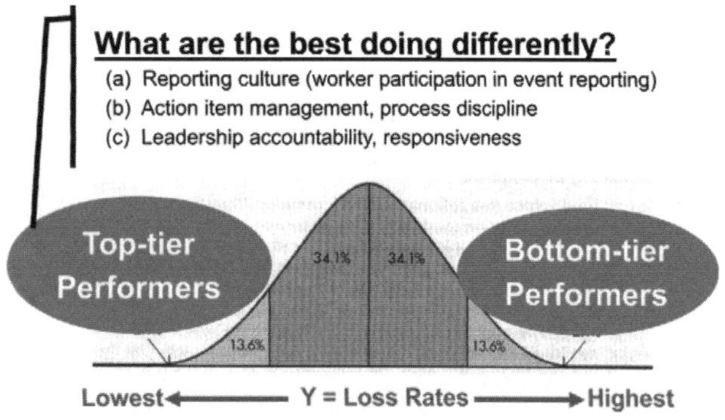

Figure 8: Finding the unique characteristics of the top performers

As conveyed in the diagram to the left, the research and benchmarking initiative has revealed a list of factors that are collectively predictive of performance and meet all of the previously listed criteria.

These factors were revealed after many repeated cycles of statistical analysis and review by SME's. The final list of the most meaningful metrics was then calculated in quartiles. From there indices were created to establish a leading performance measurement for each of the three factors. Finally the indices were combined into an overarching index to provide a single score which was found to be most predictive of WTRIR.

The Ultimate Leading Indicator: A Composite Safety Performance Index

The composite index is made up of the three dimensions: ***Reporting Culture Composite Index***, ***Action Item Composite Index*** and ***Leadership / Process Composite Index***. Each dimension is calculated from multiple metrics to form a unique measurement that results from the "tension" between the combined metrics.

Index from "tension" between metrics

For example, by combining a measure of the volume of action creation with a measure of discipline in on-time completion, there is tension between volume and disciplined execution – i.e. it is easier to complete a very short list of actions on-time than it is to carry such disciplined execution across a much longer list of action items.

Reporting Culture Composite Index = "culture of reporting and fixing"

This index results from the tension between the volume of "voluntary" event reporting and the organization's tendency to include actions as follow-up to reported events. In order to perform well in this index, an organization must not only sustain a substantial rate of near miss reporting (normalized by work hours) but also must follow-up reported events with actions. An organization must be strong in BOTH areas to be a good performer in the Reporting Culture Index.

Action Item Composite Index = "rate of action with timely execution"

This index results from the tension between the volume of actions generated and the organization's tendency to complete actions in a timely manner per the planned due date. In order to perform well in this index, an organization must BOTH sustain a relatively high rate of action AND continue to complete actions in a timely manner.

Leadership / Process Composite Index = "responsive, disciplined leadership"

This index results from the combination of supervisor / leadership involvement in two key steps in the risk reduction process lifecycle – i.e. the inital response to a reported event and the initial authorization of actions derived from events. To be a top performer in this area, an organization must have low lag times in BOTH of these areas.

The Importance of Reporting Culture

A common view of reporting culture is that it is an organization's total event reporting volume. Per the above index components, it is evident that event reporting "volume" has a critical affect on both of the first 2 above indices. Event reporting volume directly affects the Reporting Culture Index and indirectly affects the Action Item Index in that total action volume is minimized if reported events are minimized. Thus two-thirds of the total safety performance index discovered through the research project is affected by the volume of reported events – i.e. Reporting Culture.

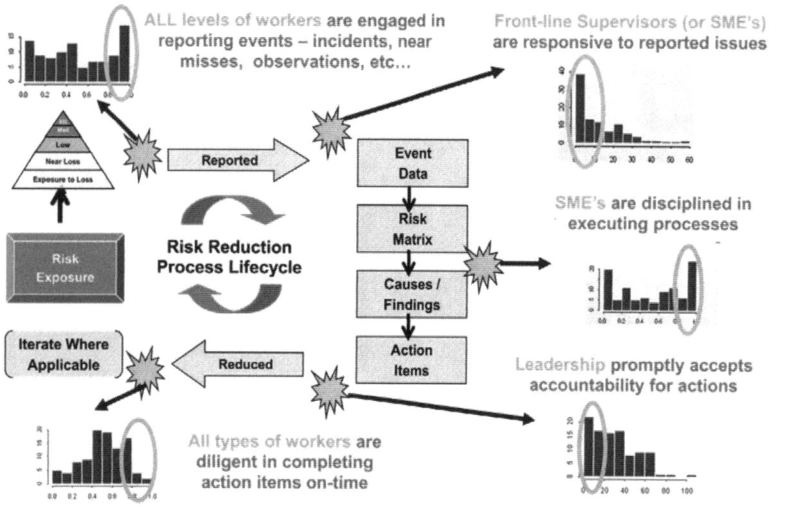

Figure 9: The unique tendencies of top performing organizations

Reading the histograms (to the right) in a clockwise manner not only portrays a list of math-grounded leading metrics but also tells a practical and believable story about organizational tendencies.

By studying the statistical relationships between the many leading metrics and WTRIR, it was revealed that organizations which perform well in each of these areas were much more likely to be top-level QHSE performers. Since these types of leading metrics are contained in the same database tool, they are applied as KPI's via the same tool to continuously improve the factors that lead to better performance outcomes.

Applying the Research Findings in Practice

Some companies have corporate databases which are justified on the ability "to produce better pareto charts of event data". But if these tools could directly DRIVE better QHSE performance, the true return on investing in these programs would be achieved. Many companies are preparing analytical reports for which some do include "meaningful" leading metrics. But are they getting the commitment of operations leaders to use those reporting tools to directly monitor and improve (i.e. measure and act upon) the math-proven factors that result in better performance? Leaderhip commitment to implement such leading KPI's as management controls is critical to drive improvements in QHSE outcome performance. Helping to get this commitment is a key purpose of the analytical research and the overall KPI measurement strategy.

Figure 10: Leadership as "Users" to continuously monitor and improve

To improve, leaders must be the KEY USERS of the continuous improvement solution.

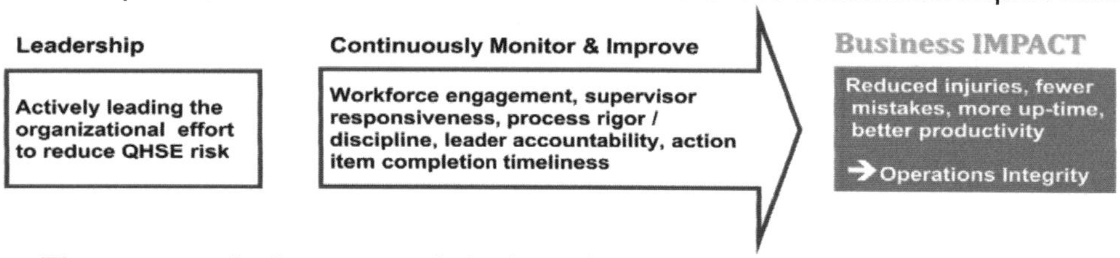

The research has revealed that the measurable math-proven keys to improvement are "organizational" in nature – i.e. workforce engagement, leadership accountability, etc.. Active management sponsorship is required to truly lead the required improvements in these areas. So, to achieve continuous improvement, it is imperative to get the data and math to help obtain management buy-in as well as the iterative benchmarking measurement tools to sustain their focus year over year.

Managed KPI Scorecard			
	Recordable Rate (Lagging)	Safety Triangle Ratio (as %)	% High Actions On Time
Div 1	2.50	37%	67%
Div 2	3.10	22%	62%
Div 3	1.70	83%	95%
Div 4	1.30	75%	77%
Div 5	0.95	93%	75%

Figure 11: Measure leading KPI's to DRIVE performance

What “leaders” measure and act upon gets improved. To obtain such active response from operations leadership, the KPI’s must be carefully selected. Only those metrics which resonate at all levels of business management are included on the “managed KPI” scorecard. An example of such broadly applicable metrics is provided to the right. The managed metrics are typically calculated on a 12 month rolling basis and cascaded through multiple levels of the organization to assure improvements are realized enterprise-wide. But these only represent a sub-set of the leading indicators which have potential to affect performance outcomes.

Executive Analysis ONLY			
Predictive Index	Report Culture	Action Execution	Leader Response
4.79	2.10	24%	11%
5.25	1.40	17%	41%
17.10	7.50	54%	57%
7.10	4.10	72%	3%
22.0	18.30	46%	21%
5.20	6.10	64%	12%
10.24	6.58	45%	24%
4.10	1.50	93%	8%
9.70	6.30	71%	31%

Figure 12: Preparing top leadership to evolve the "managed" KPI's & continuously improve

To evolve management's focus year over year, another required component entails an additional set of metrics included only for analysis by members of the top-level executive steering committee. This example (to the right) displays a sample list of the math-grounded metrics discovered through the research. Although many organizations may not quite be ready to hold managers accountable for such a list of metrics, these additional benchmark KPI's may reveal valuable trends and help to prepare the top leadership to consider such metrics for inclusion as "managed" KPI's in the subsequent measurement cycle – establishing a continuous improvement process.

Iterating with Benchmarks to Continuously Improve

The traffic light metrics are derived from benchmark calculations. Companies may choose from various types of benchmark calculations during design of the KPI reports. Internal benchmarks provide comparisons to other organizational units' performance within the same company. And external benchmarks provide comparisons to the aggregate scores per other companies using the same database tools and involved in the collective benchmarking efforts. Additionally, a discrete scale may be used to establish the traffic light scoring effect.

Division/Site	TRIR	WTRIR	Preventive Index	Reporting Culture	Action Execution	Leader Response
Org Unit 1	0.60	0.60	1	1	1	3
Org Unit 2	0.44	0.44	1	2	1	2
Org Unit 3	0.47	0.47	2	2	2	4
Org Unit 4	0.76	0.76	2	4	2	2
Org Unit 5	0.00	0.00	1	1	1	2
Org Unit 6	0.41	0.41	2	2	3	2
Org Unit 7	0.78	0.78	2	2	3	2
Org Unit 8	0.42	0.42	3	4	2	2
Org Unit 9	0.23	0.23	1	1	1	2
Org Unit 10	0.91	0.91	3	2	3	4
Org Unit 11	0.51	0.51	2	1	2	2
Org Unit 12	1.80	1.80	3	1	2	2
Org Unit 13	0.55	1.64	3	3	3	2
Org Unit 14	1.42	1.42	3	1	2	4
Org Unit 15	1.81	1.81	4	4	3	3
Org Unit 16	0.63	2.31	3	4	2	3
Org Unit 17	1.00	4.98	4	2	4	4

Figure 13: Example annual benchmark report

Each year the research members receive a benchmarking report at the operating unit level. The purpose of the report is to identify performance gaps and improvement opportunities in the areas measured by the index of leading measurements.

By updating these statistics periodically, a continuous improvement platform is established to leverage the knowledge gained across all member companies. As more companies get involved in the benchmarking program and more business leaders are engaged in improving the math-proven leading indicators, the statistical inferences are continually sharpened and the overall community of collaborating companies benefits with improved QHSE performance results.

Summation

Through applying structured statistical processes and subject matter expert interpretations to a large statistically significant data set, it has been discovered that top-performing organizations (in terms of Safety outcomes) are the ones which have their leadership and front-line workers engaged in a collective effort to reduce operational risks.

This discovery may not be a surprise but rather a validation of existing beliefs. However by deriving these facts from a system which manages daily operational activities (e.g. incidents, near misses, investigations, observations, inspections, action items, etc.), a new opportunity is afforded operations managers to continuously measure and thus improve these organizational factors and ultimately to gain control of Operational Excellence and Safety performance outcomes.

In conclusion, the research discoveries regarding the leadership and cultural dynamics of top-performing organizations are certainly not new to the Energy industry. However the ability to institute management controls which enable all levels of leadership to measurably improve these factors on a routine, daily basis is likely NEW to most companies. But for those companies who are measuring and improving broadly applicable QHSE process-based KPI's and are engaged in the collective benchmarking programs to continuously improve, this ability has become a key part of their ongoing strategy to achieve and sustain world-class QHSE performance.

(45a) Pilot and Full-Scale Application of Technologies for Mercury Control

Noah Meeks, Research & Environmental Affairs, Southern Company Services, Inc., Birmingham, AL

With the final MATS rule going into effect in 2015, electric utilities must now consider mercury control as a needed technology in their emissions train. This talk wiill cover recent full-scale injections of activated carbon and alkali injection, as well as pilot-scale studies of non-halogenated carbons or non-carbon sorbents. We will also discuss recent mixed matrix membrane pilot studies. **(Most recent data will be included in the abstract closer to the meeting)

(45b) Acid Tunneling Vs Acidizing : A Case Study of Rudatain Oilfield-Kuwait

Adel Elsharkawy, Petroleum Engineering, Kuwait University, Kuwait, Kuwait and Mohamed Al-Hadher, Pet Eng, Kuwait Oil Company, Safat, Kuwait

Acid Tunneling Vs Acidizing : A Case Study of Rudatain Oilfield-Kuwait

By Mohammed Al-Hadher and Adel Elsharkawy

Kuwait University, College of Engineering & Petroleum

Po Box 5969 Safat 13060 Kuwait

Abstract

Acidizing is one of the most important techniques used in oil industry in order to enhance the reservoirs and increase oil production. There are three types of acidizing techniques which are matrix acid stimulation, acid fracturing, and acid tunneling. In this paper a comparison these types of acidizing is discussed. The case study of RA-0007 acid tunneling completed in Kuwait Oil Company is presented. It represents the first dendritic with extensive worm holes created by acid tunneling for maximum reservoir contact. The whole effect is aptly demonstrated by the change in PI of the well before and after the acid tunneling as mentioned before. Analysis of the well designed and implemented job showed that two tunnels were created successfully in MA-C formation through a two days rig-less operation. The results indicated an improvement in the well productivity index (PI) which had been increased from 1.3 to 3.87 bpd/Psi.

(45e) Sustained Casing Pressure Modeling and Well Integrity

Tony Rocha-Valadez, Mary Kay O'Connor Process Safety Center, Artie McFerrin Department of Chemical Engineering, Texas A&M University, College Station, TX, A. Rashid Hasan, Harold Vance Department of Petroleum Engineering, Texas A&M University, College Station, TX, C. Shah Kabir, EPT, Hess Corporation, Houston, TX and Sam M. Mannan, Artie McFerrin Department of Chemical Engineering, Mary Kay O'Connor Process Safety Center, Texas A&M University, College Station, TX

SUSTAINED CASING PRESSURE MODELING AND WELL INTEGRITY

T. Rocha-Valadez[1*], A.R. Hasan[2], C.S. Kabir [3] and S.M. Mannan[1]

*[1]Mary Kay O'Connor Process Safety Center, Artie McFerrin Department of Chemical Engineering, Texas A&M University, College Station, Texas 77843-3122, US., email: *tonyrochav@tamu.edu, mannan@tamu.edu.*

[2] Harold Vance Department of Petroleum Engineering, Texas A&M University, College Station, Texas 77843-3122, US, email: rashid.hasan@pe.tamu.edu.

[3]Hess Corporation, 1501 McKinney Street, Houston, TX 77010, US, email: skabir@hess.com.

With an increasing demand for hydrocarbons and other sources of energy, the methods for delivering such energy needs have become complex. In the case of the oil and gas industry, exploration and production have gone to deeper waters and deeper horizons than ever before, resulting in drilling several miles into the earth's core. Deeper well penetration means encountering increasingly high formation pressures and temperatures. For example, in deepwater drilling, well construction and production management deal with significant heat transfer between the produced fluids and its surroundings. The high-rate production and the attendant heat transfer demand close attention in terms of operational safety and potential environmental impact. One phenomenon that occurs in producing wells is sustained casing pressure (SCP), which presents a risk to both process safety and the environment.

In oilfield terminology, the casing is a steel pipe cemented to the surrounding formation. Multiple concentric casings are involved in a typical well construction. The casing immediately surrounding the production tubing is known as the production casing, which is most vulnerable to thermal and mechanical stresses and fluid leakage. The cement sheaths provide zonal isolation, support the casing, and protect it from corrosive fluids. Ideally, pressures monitored at the wellhead for multiple casings should change minimally after being bled off through a needle valve at the surface. When the pressure rebuilds after being bled off, that casing is said to exhibit SCP [1].

SCP occurs because of stresses that the system experiences owing to fluids production up the tubing. Besides the normal source of SCP, a poor cement job may allow hydrocarbon migration from the producing or other horizons and accumulation into one or more of the casings. The risk of not attaining well integrity can range from the activation of rupture discs to the blowout of hydrocarbons. Because of the safety and environmental dangers of SCP, regulations require remedial treatments for a casing string that exhibits significant SCP problems. Unfortunately, available methodologies to test for SCP are often qualitative and empirical in nature [2, 3], rendering them inapplicable in many situations. No practical systematic testing procedure has emerged to quantify SCP.

In this paper, we present a model for SCP that is rooted in the transport processes of the system. The model assumes gas leaking through the cement barrier, and then migrating up the annular-liquid column. It accounts for the fluid-temperature profile along the wellbore and liquid compressibility. Some robust assumptions about the transport processes allow us to write the governing equation as a first-order, linear-differential equation that is solved with appropriate boundary conditions. The resulting algebraic expression for the casing pressure rise as a function of time paves the way for all applications. A comparative study of the model's performance with field data suggests excellent agreement for wells where the casing pressure increase is owes gas influx into the annulus. The proposed model could form the basis for standardization of SCP testing. This study also offers some test procedures to assess SCP-related safety concerns.

References

1. Bourgoyne Jr., A.T.J., S.L. Scott, and W. Manowski, ***A Review of Sustained Casing Pressure(SCP) Ocurring on the OCS***, 2000, Final Report Submiteted to MMS.

2. Xu, R. and A.K. Wojtanowicz, ***Diagnosis of Sustained Casing Pressure from Bleed-off/Buildup Testing Patterns.*** SPE, 2001: p. Productions and Operations Symposium.

3. Wojtanowicz, A.K., S. Nishikawa, and R. Xu, ***Diagnosis and Remediation of Sustained Casing Pressure in Wells***, July 2001, MMS.

(45f) Evaluation of Particle Growth in an Electrochemical Fluidised Bed Reactor

Emmanuel Ehirim, Rivers State University of Science and Technology, Port Harcourt, Nigeria

The process of electrodeposition within electrochemical reactor results in the increase of particle sizes, which increases the volume of the bed. This work entails a mathematical model to describe the effect of this phenomenon on the behaviour of a three-dimensional fluidised bed electrode operating under the condition of electron transfer. The model incorporates the pseudo-stationary behaviour in the reactor since the bed increases in volume due to the particle size growth. The Orthogonal Collocation method was used to solve the developed differential equations. The overpotential distribution and the reaction rate within the reactor have been calculated. The principal results are the effects of the electron transfer operation on the superficial displacement with time. The effects of total applied current, electrode geometry and the bed porosity on the behaviour of the system have been analysed. The result shows a very gradual increase in the bed height.

(45g) Debottlenecking Gas Dryers and Desiccant Adsorbers

Benjamin A. Schmitt, Hydrocarbon Processing Industry (HPI) - Business Unit, Johnson Screens, Inc., New Brighton, MN and Michael Ekholm, Engineering, Johnson Screens, Inc., New Brighton, MN

Debottlenecking Gas Dryers and Desiccant Adsorbers

Though Maximization of the Useable Bed Volume with Johnson Screens' SSG™

Benjamin Schmitt, PE

Senior Mechanical Engineer, Technology

Michael Ekholm, PE

Applications and Product Design Manager

Johnson Screens Inc. St. Paul, MN

AIChE Spring Meeting, April 2013

13th Topical Conference

San Antonio, TX

Introduction:

The following presentation of the Shaped Support Grid (SSG™) releases additional supporting data and updates to the initial presentation given by Robert G. Norell at the 2010 AIChE Spring Conference.

Limitations on ethylene dryers and desiccant adsorbers can create bottlenecks in the production process which can lead to a need for increased dryer and adsorber capacity. Capital intensive investments may be necessary in order to eliminate the bottleneck which may include the addition of an additional vessel. With limits on the plot plan, this may not be a workable solution. Adding to the problem is the amount of time consumed in engineering, planning and construction, which can have negative effects on the operation of the ethylene unit. Increased desiccant volume and operational efficiency can increase the unit's production with minimal capital expenditures. Johnson Screens', patent pending, SSG™ can add the increased process efficiency and cycle life necessary to extend the operating capability of the unit.

Extending the operation, though debottlenecking, can be achieved by maximizing the time before breakthrough of the media bed. There are three primary functions that maximize the breakthrough time:

Proper gas distribution in both normal operation and regeneration conditions which maximizes the bed utilization.

Increased molecular sieve or desiccant bed volume are achieved through the elimination of traditional bed supports comprised of support grids and beams. The support interface for the screen configuration also allows for thermal expansion typically present in vessels with a thermal cycle as part of the process. The design allows for thermal cycling without the risks of seal or structural failure inherent in other solutions.

Minimizing the bed volume consumed by the distribution and support system increases the net desiccant volume lost when utilizing alternative bed supports.

Johnson Screens ***SSG*** is designed to provide the industry with an alternative, cost effective, approach to increasing capacity. Supporting data and preliminary case studies will be depicted to quantify the benefits that are achievable through the implementation of the ***SSG***.

Conclusions:

The use of a Johnson Screens ***SSG*** provides a reliable immediate solution to reduce capacity bottlenecks. The ***SSG*** increases the bed volume, when compared to traditional bed support arrangements comprised of grids and beams. The increase in desiccant volume will allow for increases in feed flow rate and/or adsorbtion cycle time. The ***SSG*** maximizes the flow distribution during desiccant regeneration, maximizing the utilization of the complete cross section of the vessel. This solution can be implemented in both new and existing vessels, providing consistency in design between planned increases in performance of current vessels and future expansions.

The modular design of the ***SSG*** allows for easy installation and maintenance. Connection of the ***SSG*** to the vessel is performed within the outlet nozzle, eliminating all welding on the pressure vessel and eliminating any need for recertification of the pressure vessel after installation. The design allows for ease of dismantling for inspection of the vessel head during routine maintenance and inspections. The low profile will also reduce interference with any required media exchanges. The design package of the ***SSG*** provides the end user with a holistic solution that fulfills debottlenecking needs with limited capital expenditures.

Biography of Speaker:

Benjamin Schmitt is a Senior Mechanical Engineer in Johnson Screens' Hydrocarbon Processing Industry Technology Group. Ben graduated from North Dakota State University in 2004 with a Bachelor of Science degree in Mechanical Engineering. He currently working toward his Master's in Business Administration from Hamline University. Ben is member of AIChE and is a registered professional engineer in the state of Minnesota.

AMACS
PROCESS TOWER INTERNALS

AIChE 2013 Spring Meeting Poster board
Design Considerations for effective liquid removal In Suction Drums

Presented By:
Kanti Patel, PE

Compressor down again?

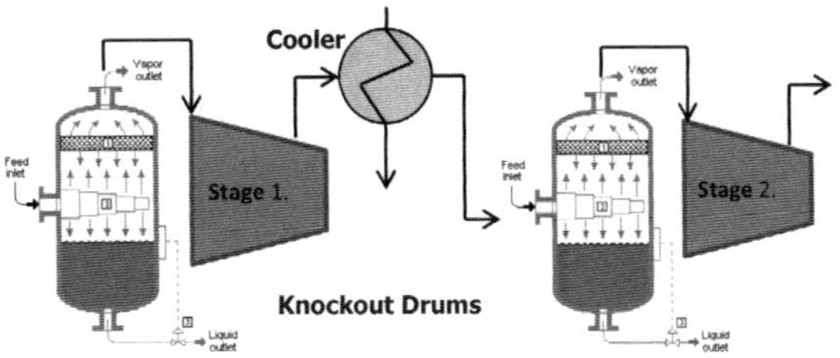

Is liquid carryover the culprit?

<u>We have heard it all before</u>:

-Premature compressor shutdown.

-Production time lost.

-High yearly compressor maintenance costs.

-Impeller damage due to pitting corrosion.

-Corrosion due to salt deposits.

AMACS
PROCESS TOWER INTERNALS

Common Design Problems with Suction Drums which promotes liquid carryover

1. Choosing the correct mist eliminator (M.E. equipment)
2. Knowing where to install the M.E. in a suction drum.
3. Consider a high mist loading in the gas when selecting your M.E.
4. Fouling considerations.
5. M.E. disengaging space, bad gas distribution & liquid slug in the feed.
6. One micron and great droplets, versus submicron mist removal.
7. High viscosity liquids.
8. Dislocated M.E. due to pressure surges.
9. Through put exceeds original design capacity.
10. Capacity range too large for suction drum.

AMACS
PROCESS TOWER INTERNALS

Choosing the correct Mist eliminator equipment

- Proper sizing

+ Use Souders-Brown equation to find optimum velocity.

$$\mathbf{Vd = K} \cdot \sqrt{\frac{Pl}{Pg} - 1}$$

+ Select the correct vapor velocity factor K. based on the operating pressure, and type of equipment.
  - Lower K-factor 0.01 for every 100 psi
  - K factor to select when using:

Mesh Pad	K= 0.1 - 0.55
Chevron Vane	K= 0.45 - 0.65
MultiPocket® Vane	K= 0.75 – 1.0

AMACS
PROCESS TOWER INTERNALS

Choosing the correct Mist eliminator equipment cont.

- Selection based on droplet size.

Table 2. Droplet sizes (water in air) typically captured with 99.9% efficiency by mist eliminator elements of various types

Element type	Size range microns
Fiber candles or panels	0.1 and larger
Mesh with co-knit yarn	2.0 μm and larger
0.006-inch knitted mesh	5.0 μm and larger
0.011-inch knitted mesh	10 μm and larger
AMACS MultiPocket ® Vane	10 μm and larger
Conventional vane arrays	15 μm and larger

AMACS PROCESS TOWER INTERNALS

Where to install the M.E. in the suction drum

- Improperly installed M.E. can cause severe liquid carryover problems.
- General accepted spacing guidelines.

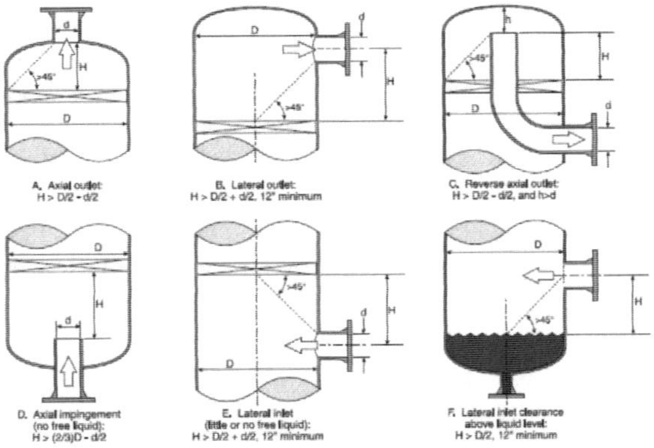

AMACS PROCESS TOWER INTERNALS

Consider high mist loading when selecting your M.E. equipment

+ High mist loading in the gas stream can flood your M.E. equipment and cause liquid carryover.

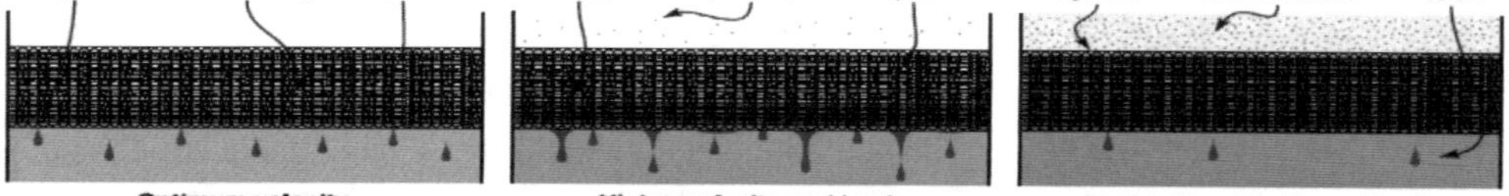

Figure 4. Envisioning mesh pad performance degradation as liquid and gas loads increase (vertical cross-sections)

AMACS
PROCESS TOWER INTERNALS

Consider high mist loading when selecting your M.E. equipment cont.

+ Limited drainage capabilities
  - mesh M.E. <0.5 gal/FT2/min
  - Vane type M.E. <2 gal/FT2/min

+ Select a combination of vane & mesh mist eliminators

AMACS
PROCESS TOWER INTERNALS

Fouling Applications

+ Fouling of mesh mist eliminator will cause liquid carryover.

+ For mild fouling install a water spray system below the mist eliminator for occasional inline cleaning (based on Delta-P)

+ For consistent fouling problems, install a Vane mist eliminator upstream of the mesh mist eliminator.
  - Vanes remove solids and bulk liquids therefore, protecting the downstream mesh mist eliminator

AMACS
PROCESS TOWER INTERNALS

Fouling Applications cont.

+ For consistent fouling problems where the solids are "sticky" install a water spray system below the vane mist eliminator.

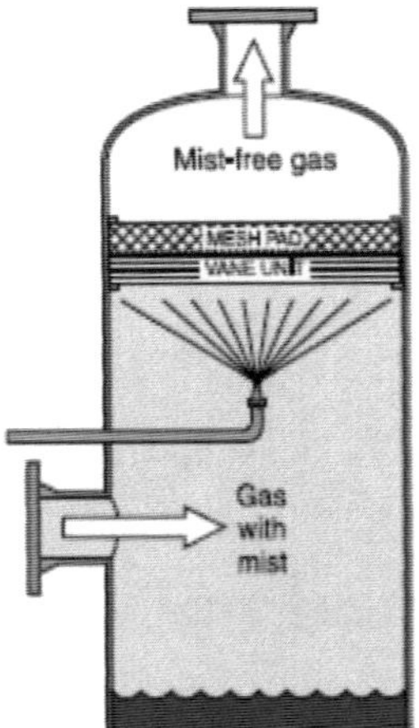

AMACS
PROCESS TOWER INTERNALS

Bad Gas distribution, limited disengaging space and liquid slugs.

+ Bad gas distribution, limited disengaging space in a suction drum can cause severe liquid carryover.

+ CFD analysis can assist in locating high velocity areas.

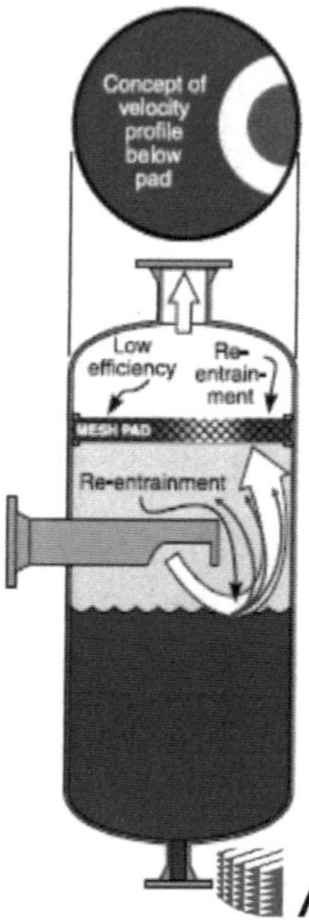

AMACS
PROCESS TOWER INTERNALS

Bad Gas distribution, limited disengaging space and liquid slugs cont.

+ Use inlet diffuser to force uniform gas distribution i suction drum.

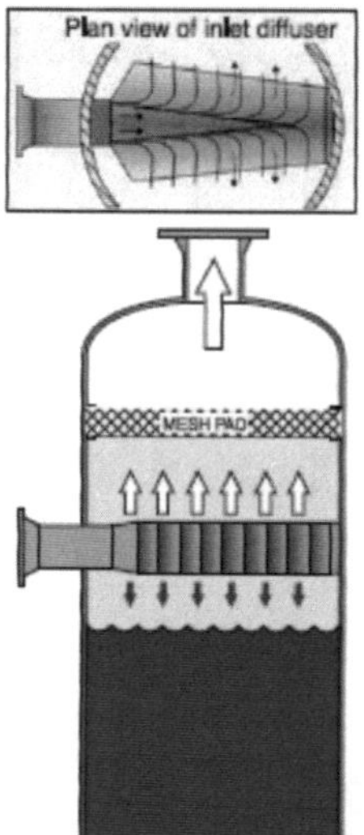

AMACS
PROCESS TOWER INTERNALS

Inlet Feed Device

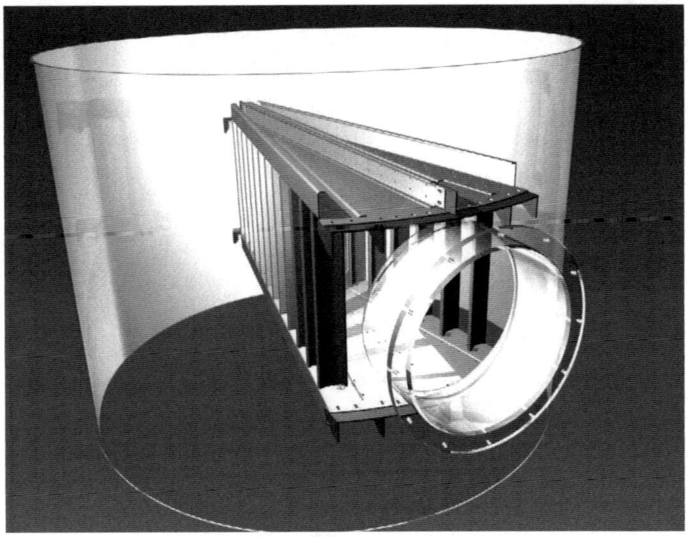

AMACS
PROCESS TOWER INTERNALS

Inlet Feed Device

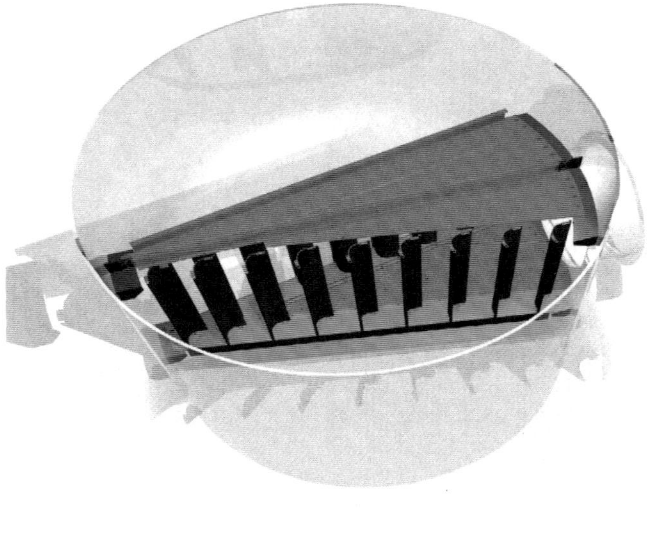

AMACS
PROCESS TOWER INTERNALS

Bad Gas distribution, limited disengaging space and liquid slugs cont.

+ Use a perforated distribution plate downstream of the M.E. for additional distribution assistance.

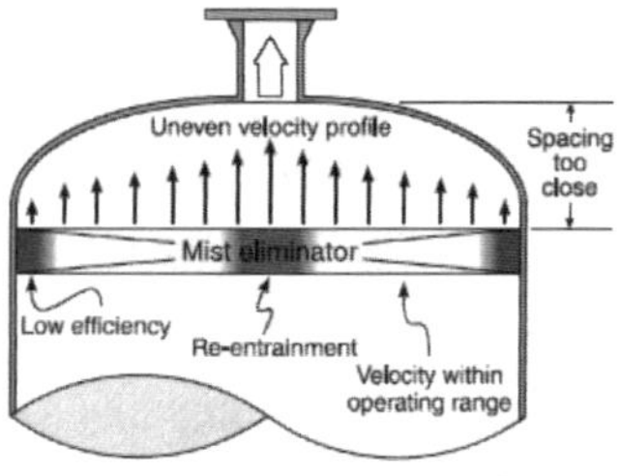

Figure 7. Example of mist eliminator performance degradation due to uneven velocity profile

+ Inlet diffusers also handle incoming liquid slug which otherwise can cause flooding of the Mist eliminator.

AMACS
PROCESS TOWER INTERNALS

Know what mist size droplets You are dealing with

+ Know if your mist is > 1 micron or submicron.
+ When gas travels from A to B, condensation in the pipeline will cause a fine mist (fog or submicron droplets)

  Examples:

  - Incoming gas from an offsite location
  - Fuel gas to furnaces/heaters with low NOx burners

AMACS
PROCESS TOWER INTERNALS

Know what mist size droplets You are dealing with cont.

+ Submicron mist is not captured by conventional M.E.

+ Submicron mist requires a coalescing filter, typically
installed in a two stage filter separator.

AMACS
PROCESS TOWER INTERNALS

Know what mist size droplets You are dealing with cont.

<u>1^c stage;</u> submicron droplets are coalesced to > 10 micron droplets.

<u>2^c stage;</u> greater than > 10 micron droplets are ejected from the coalescing filter and captured by a vane, which captures droplets > 8 micron.

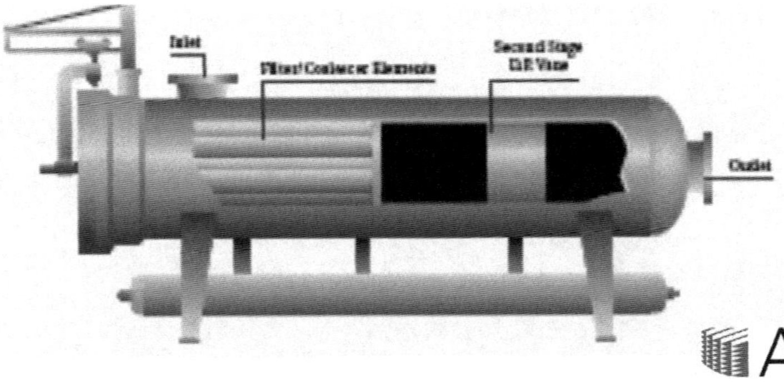

AMACS
PROCESS TOWER INTERNALS

High viscosity liquid as mist

+ Viscous liquids drain slower out of the M.E. and could cause localized flooding.
+ Use a mesh mist eliminator with an open drainage layer on the bottom to improve liquid drainage.
+ Use a lower K- value (lower face viscosity) for the M.E. To allow additional drainage time.
+ Use a vane type M.E. upstream of the mesh mist eliminator, to remove the bulk of the viscous liquid, so the M.E. can effectively handle the residual liquids.

Dislocated M.E. due to System pressure surges

+ M.E. sections can be dislodged into the downstream equipment, which then creates massive carryover.
+ Standard M.E. equipment often gets installed and secured to the support ring with tie wire.

Dislocated M.E. due to System pressure surges cont.

+ Where system pressure surges are common, there are more secure ways to install the M.E.
  - dual support rings
  - through bolts
  - install a pressure relief door

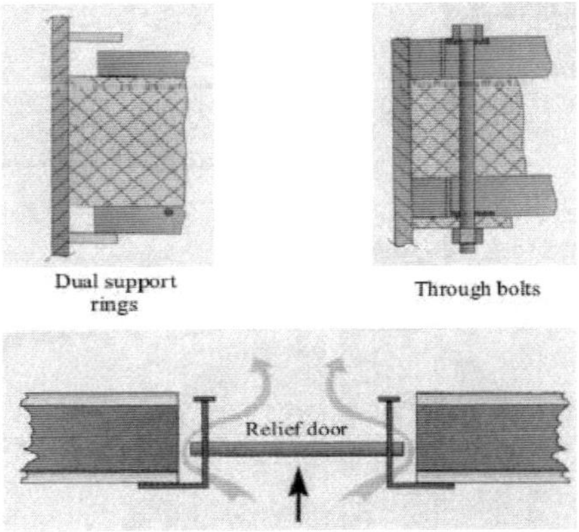

AMACS
PROCESS TOWER INTERNALS

When throughput exceeds original capacity

+ There are various solutions to increase capacity in an existing vessel.
  - Patented High capacit
    mist eliminator 0% - 1(

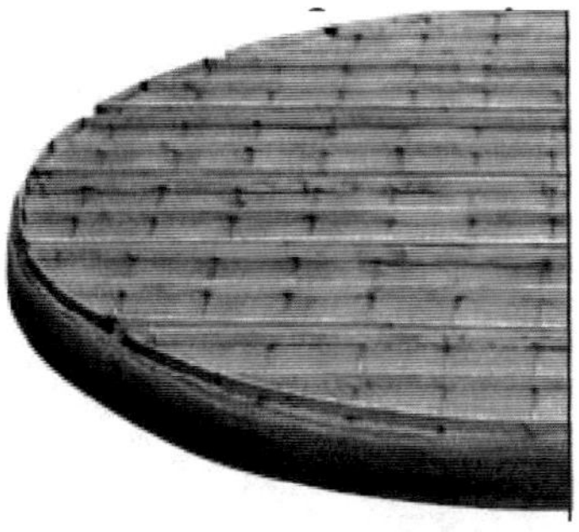

AMACS
PROCESS TOWER INTERNALS

When throughput exceeds original capacity cont.

-Patented MISTFIX™ cylindrical mist eliminator 0% - 25%

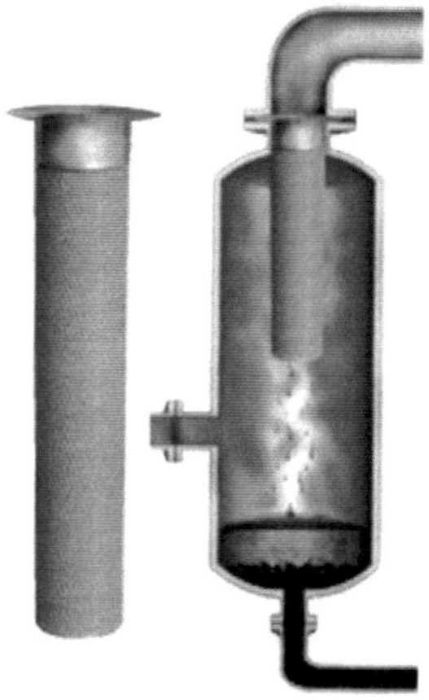

When throughput exceeds original capacity cont.

- High capacity Patented MultiPocket® vane. 0%-50%

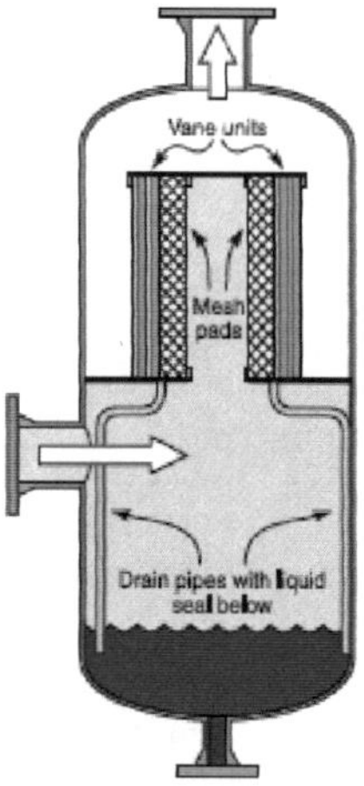

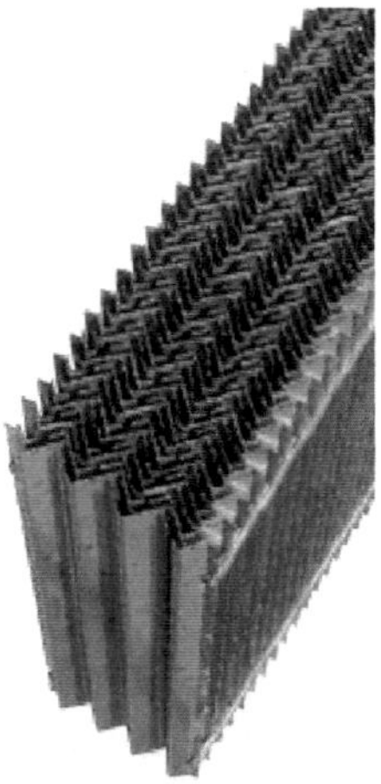

AMACS
PROCESS TOWER INTERNALS

When throughput exceeds original capacity cont.

+ MistFix™ is installed through the gas outlet, all other M.E. equipment comes in sections, installed through the manway.
+ With increased capacity comes increased bad gas distribution.
  - Use an inlet diffuser.
+ To avoid welding new support rings and housing for the high capacity new equipment, use double expansion rings, a compression fit.

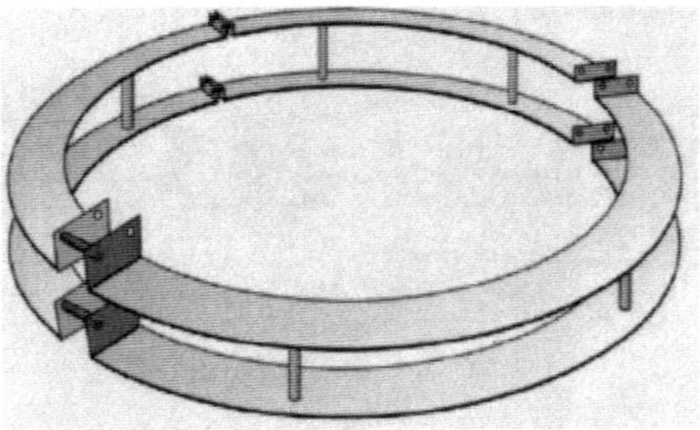

AMACS
PROCESS TOWER INTERNALS

Support Rings

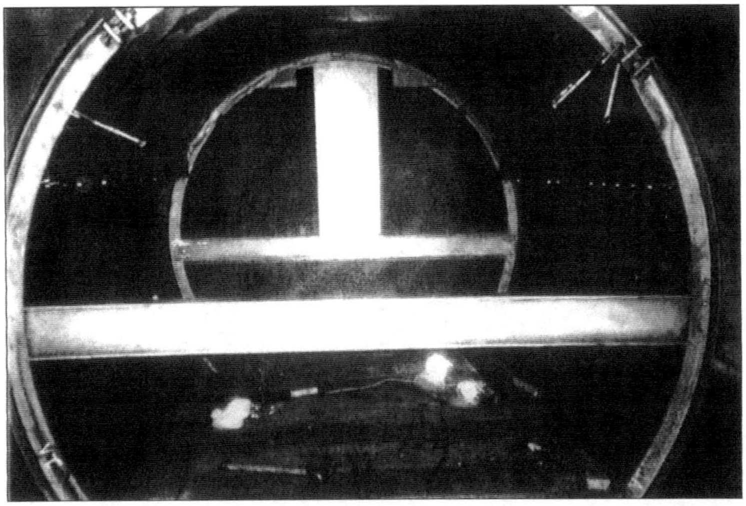

AMACS
PROCESS TOWER INTERNALS

Support Rings

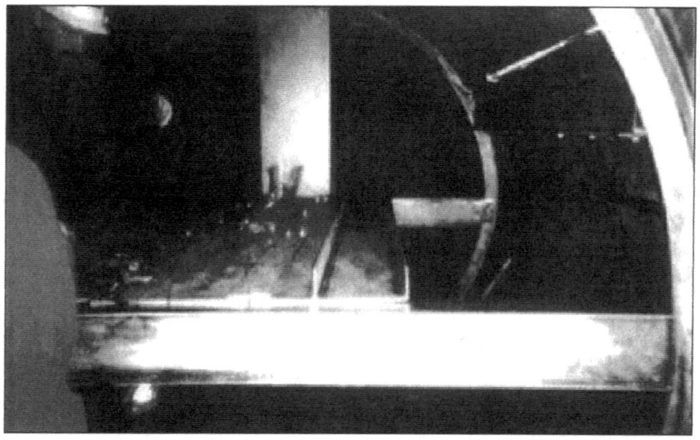

AMACS
PROCESS TOWER INTERNALS

When the capacity range is too large for a suction drum

+ Typical M.E. equipment will operate between
  30% – 110% capacity.

+ When a suction drum often operates below 30% liquid carryover may occur.

AMACS
PROCESS TOWER INTERNALS

When the capacity range is too large for a suction drum Cont.

To facilitate a capacity range of 10% - 110% consider a 2 phase M.E. in the same drum.

- one that operates 30% - 1
- second that operates between 10% - 30%

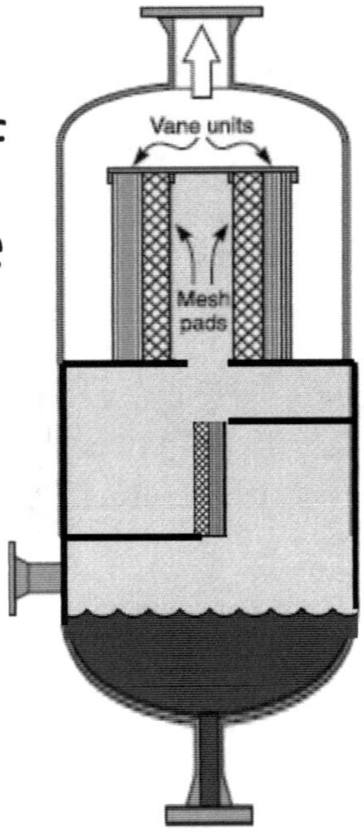

(45i) Renewable Feedstock for Steam Crackers: Catalytic Upgrading of Crude Tall Oil (CTO) Into Bio-Naphtha

Jinto Anthonykutty, Juha Linnekoski, Antero Laitinen and Ali Harlin, VTT Technical Research Centre of Finland, Espoo, Finland

Renewable feedstocks from bio resources are better alternatives to petroleum derived naphtha for the production of olefins and aromatics by steam cracking or reforming.[1, 2] Mostly, these renewable feedstocks (bio-naphtha) rich in n-alkanes are produced by catalytic upgrading methods via hydrotreating/hydroprocessing; in which oxygenates are removed from bio-oil by various deoxygenation processes. At present, hydrotreated vegetable (HVO) oil is produced in many countries on industrial scale using existing petroleum refinery methods[3]. However, employing higher priced vegetable oils could be a major concern in these processes and the main challenge is therefore finding an inexpensive and non-food chain affecting feedstock. Tall oil, obtained as a by-product from Kraft pulping process, is abundant and cheap;and hence, meets the criteria of a low cost starting material for upgrading[4]. Moreover, the utilization of tall oil in conventional oil refineries could serve as an eco-friendly method to convert bio-wastes into bio-naphtha[1].

In the present research approach, catalytic upgrading (hydrotreating) of crude tall oil (CTO) into bio-naphtha suitable for a steam cracker is discussed. CTO, which contains mainly fatty acids (42.9 wt. %) and resin acids (25.7 wt. %) was obtained from Stora-Enso Kaukopää (Finland) mill and used as received. A commercial NiMo catalyst was employed and the experiments were conducted in a continuous, down flow, fixed bed reactor at different process conditions (WHSV 1-3h^{-1}, Pressure 5 MPa and Temperature 350 - 450 ^{o}C). After the stabilization period (6 h.), the liquid sample was collected and separated into organic and aqueous phases.The product distribution in organic phase was analyzed with a Shimadzu GC/MS equipped with a Zebron MS 30m × 0.25 mm × 0.25μm column.

This current approach reports that CTO produces mainly paraffins, cycloalkanes and aromatics during upgrading process. Paraffinic hydrocarbons mainly comprise n-octadecane (20-25 wt. %) and n-heptadecane (15-20 wt. %), are obtained in maximum yield (55- 60 wt. %) in low temperature reactions. The yield of paraffinic hydrocarbons increased as a function of increasing space time in low temperature reactions. 18-Norabietane and other mono-, di- and tri- cycloalkanes are obtained from resin acids in CTO, and these cyclic structures are converted to aromatics such as naphthalene and phenanthrene in high temperature reactions (>400 ^{o}C). Stigmastane, apparently produced from sterols in CTO are also present as a major three ring component in the product stream. Additionally, gaseous products obtained and the yield increased as a function of temperature. The trend observed with the distribution of gaseous products reveal that decarboxylation and cracking reactions are favored at higher temperatures, while hydrogenation and hydrodeoxygenation (HDO) are suppressed.

The elemental composition of CTO shows that it contains relatively high sulfur content (1800 mg/kg), could be useful to keep the catalyst's activity with time on stream (TOS). However, the considerable amount (87 mg/kg) of alkali metals (Na, K, Ca and P) in CTO might be critical for the catalyst activity with TOS, will be discussed in this contribution.

The product composition obtained with CTO is assessed in comparison with compositions and characteristics of refinery naphtha feeds, and will be discussed in detail. Furthermore, this contribution will discuss the achieved deoxygenation rate and success criteria for the product composition of upgraded CTO.

References

1. Gornay, J.; Coniglio, L.; Billaud, F.; Wild, G. Steam Cracking and Steam Reforming of Waste Cooking Oil in a Tubular Stainless Steel Reactor with Wall Effects. ***energy&fuels***, **2009**, 23(11), 5663-5676.
2. Pyl, S.P.; Schietekat, C.M.; Reyniers, M.F.; Abhari, R.; Marin, G.B.; Van Geem, K.M. Biomass to olefins: Cracking of renewable naphtha. ***Chemical Engineering Journal***, **2011**, 176-177, 178-187.
3. Sunde, K.; Brekke, A.; Solberg, B. Environmental Impacts and Costs of Hydrotreated Vegetable Oils, Transesterified Lipids and Woody BTL- A Review. ***Energies***, **2011**, 4(6), 845-877.
4. Coll, R.; Udas, S.; Jacoby, W.A. Conversion of the Rosin Acid Fraction of Crude Tall Oil into Fuels and Chemicals. ***energy&fuels***, **2001**, 15, 1166-1172.

AIChE Paper Number 45j
High Efficiency Ethylene Cracking Furnaces

Rajaram Ramesh

Product Development Manager - Ethylene
Technip
Netherlands

Marco W. M. van Goethem, PhD

Product Development Manager – Modeling and Testing
Technip
Netherlands

25th Ethylene Producers Conference
AIChE 2013 Spring National Meeting
San Antonio, Texas, April 29 – May 3, 2013

Abstract

Environmental and economic pressures generate a driving force to produce products more efficiently. A direct effect of more efficient processes is that the emission of green house gases, particularly carbon dioxide decreases. In households nowadays "high efficiency furnaces" are available with efficiencies in the range of 94-97%. This paper describes the outcome of a feasibility study on applying the high efficiency furnace concept to steam cracking furnaces. The high efficiency is achieved by enhanced heat recovery from the flue gases before they are released to the atmosphere at ca. 65°C.

Technip studied the application of this concept for cracking furnaces based on gas as well as liquid feedstocks. Natural gas was used as basis for the energy price. Since energy prices have a strong impact on project economics, the evaluations were done for an EU average natural gas price and a US natural gas price.

Two alternative heat sinks have been considered: preheating of combustion air and preheating of cold hydrocarbon feed. Both are viable for liquid cracking furnaces. In the case of gas cracking furnaces feed preheating results in a pinch and this makes it not viable.

Capital costs are based on Technip's cost database and quotations from air preheater vendors. Special attention has been paid to the material selection of the affected convection section banks.

This study shows that high efficiency ethylene cracking furnaces are economically viable. For Europe the study indicates an attractive payback period. In the US the payback period is longer due to the lower energy prices. If energy prices increase in the future, the concept's economic viability will become better. Inclusion of CO2 credits will make the payback period more attractive.

Furnace Efficiency

In this paper furnace efficiency has been calculated as net thermal efficiency as interpreted by API560[1)]. Furnace efficiency is determined by drawing an envelope around the furnace and including in it the radiant section, convection section and any heat exchanger equipment that recover heat from the flue gas (air preheater, economizer, steam superheater, etc.). The efficiency is the ratio of the useful heat absorbed to the fired heat of the fuel and any other external heat inputs. Ambient temperature is a reference temperature for efficiency calculations and any heat inflow or outflow above this reference is taken into consideration as a heat input or heat loss.
Furnace efficiency can be determined as follows:

$$\frac{heat\ absorption\ in\ radiant\ section\ +\ convection\ section}{(firing\ rate\ \times\ fuel\ heating\ value) + sensible\ heat\ of\ fuel\ and\ air\ above\ ambient\ temp. + other\ external\ heat\ inputs}$$

Alternately, Furnace efficiency $= 100\% -$ Total heat loss, where:

$$\text{Total heat loss} = \frac{stack\ loss\ +\ radiation\ heat\ loss}{(firing\ rate\ \times\ fuel\ heating\ value) + sensible\ heat\ of\ fuel\ and\ air\ above\ ambient\ temp. + any\ external\ heat\ inputs}$$

In a furnace where fuel and combustion air enter the furnace at ambient temperature and there are no other external heat inputs, the first expression can be simplified to the following:

$$Furnace\ efficiency\ = \frac{heat\ absorption\ in\ radiant\ section\ +\ convection\ section}{fuel\ firing\ rate\ \times\ fuel\ heating\ value}$$

Please note that net thermal efficiency, as defined by API 560, is based on the lower heating value of the fuel.

Study objectives

Technip undertook this study with the goal to reduce fuel firing in cracking furnaces by at least two percent. Although this figure does not look very ambitious, it must be remembered that since efficiencies in current designs are already in the mid nineties (upper part of the technology S-curve), further improvements will not be easy and can only be incremental.

Since stack loss is the main source of inefficiency (radiation heat loss is comparatively smaller and more difficult / expensive to reduce) it was decided to make stack loss reduction as the main focus of the study.

By a relatively straight forward calculation it can be shown that the two percent efficiency rise target would require the flue gas to be cooled to approximately sixty or seventy degrees Celsius.

The main challenge in reducing stack temperature to this level is finding an appropriate heat sink for the recovered heat. In a cracking furnace there are two suitable heat sinks, combustion air and hydrocarbon feed, and both were studied. The following sections discuss the results for liquid furnaces and gas furnaces.

Basis of Study

We have done the study on a naphtha furnace and an ethane furnace. We have chosen these two feedstocks since a large number of furnaces operate on these two feeds and we can draw conclusions that can be applied approximately to other liquid and gas feeds. The study has been performed on the following bases:

- The reference liquid and gas furnace designs used in this study are based on actual furnaces from Technip's past projects. The furnaces were simulated using Technip's proprietary tool SPYRO which is used in the design and simulation of cracking furnaces.
- The budgetary capital cost need to achieve the target furnace efficiency was calculated as an incremental cost. It was estimated from vendor quotes and Technip's internal cost database.
- The cost estimates are for a new project, not a revamp, and are for a West European location.
- Operating costs are based on cost of base utilities and derived utilities. Base utilities are natural gas and demineralized water. These are taken from market data.
- Since natural gas prices are very different in different parts of the world, the operating costs and economic analysis have been done for European and United States gas prices.
- EU natural gas price was fixed at €25.1 /MWh of higher heating value (HHV) which is equivalent to €28.0 /MWh of lower heating value (LHV). The prices are based on EUROSTAT data for 1st half of 2010 and were obtained from the publications of the Department of Energy & Climate Change in the UK. They correspond to the EU-15 average for Large Industrial Consumers (i.e. 2.13 pence/kWh HHV).
- US Natural gas prices were obtained from the US Department of Energy. The average value was 6 \$/MM BTU (HHV) or 17 €/MWh (LHV) for 1st half of 2010. The average HHV used is 1029 BTU per cubic foot or 1.029 MM BTU per thousand cubic feet.
- Cost of demineralized water was assumed at € 2.00 /ton.
- The prices of other utilities (boiler feed water, steam and electricity) have been calculated from those of the two base utilities. Electricity was priced assuming 35% overall efficiency of power generation.
- The economic calculations do not include any CO2 credit for the fuel saved.
- Assumed exchange rate: 1 US dollar = 0.75 euros
- The plant is assumed to operate 8400 hours per year

The resultant utility prices used in the study are summarized below.

UTILITY COSTS

		EU basis	US basis	
FUEL as NATURAL GAS	Base	**28.0**	**17.0**	€ / MWh
DEMINERALISED WATER	Base	2.0	1.5	€ / ton
VHP STEAM (net value)	Derived	23.7	14.4	€ / ton
MP STEAM (net value)	Derived	13.9	8.5	€ / ton
LP STEAM	Derived	9.7	5.9	€ / ton
ELECTRICAL POWER	Derived	80.1	48.6	€ / MWh

Study Results

Naphtha Furnace using Air Preheat

Combustion air can be preheated in a plate type air preheater (APH). These are cross flow type exchangers and have low pressure drops. Combustion air is at a lower temperature than any of the process or utilities streams entering the convection section; therefore, from the standpoint of efficiency, APHs are ideally suited to be the last heat recovery section before the stack.

Fig 1 shows the Base Case Naphtha furnace while Fig 3 shows the same furnace with an APH. Please also refer to Table 1 for a performance comparison of the APH design with the Base Case naphtha furnace. The following points are of interest:

1) The overall efficiency of the furnace will improve from 94.4 to 97.0% for the APH option. This is mainly due to the lower stack temperature but also partly due to the smaller excess air, which is explained below.

2) This option will require a forced draft combustion air fan to overcome the pressure drop in the APH. As a result the furnace operation will change from induced draft mode to balanced draft. The availability of air in forced draft will enable more accurate control of the excess air and it will be possible to operate the furnace with lower excess air (5% instead of 8%). This leads to a lower flue gas flow, a lower stack loss and therefore better efficiency.

3) There will be a drop of 4.8% in fuel firing which is greater than the efficiency increase. This is because of the lower absorbed heat. The furnace is designed and controlled in such a way as to supply all the required process heat and to absorb the surplus heat into utilities to the extent possible. When the furnace operates more efficiently and the heat demanded by the process media remains unchanged, as in the APH case, there will be less surplus heat available for VHP steam. As a result the VHP steam generation will go down from 79.7 to 76.0 t/h.

4) The heat transfer surface in some parts of the APH will be below the water dew point of the flue gas (around 60°C). This can lead to water condensation and create two possible issues: 1) accumulation of water in the APH and 2) corrosion due to the presence of liquid water, oxygen and carbon dioxide.

   Two measures have been included in the design to counter these problems:
   - Placement of unit such that flue gas flows downward to aid easy draining
   - Use of corten steel to resist corrosion

5) We have evaluated the impact of the air preheat option on NOx emissions using API 535[2)]. The rise in air temperature and the drop in excess air will have opposing effects on the NOx production. The net effect is that the NOx concentration will be close to the Base Case.

   If a DeNOx unit is present, as is often the case in the US, the APH option may help to slightly extend the life of the DeNOx catalyst. This is because the total NOx load will be smaller than the Base Case due to the smaller flue gas flow.

Table 2 summarizes the results of the economic analysis of this option. We have used payback period to compare different options. Payback period is the ratio of the incremental capital cost above the Base Case CAPEX divided by the savings in yearly operating cost.

1) The capital cost (CAPEX) of the APH option includes the cost of the APH, the combustion air fan, extra ducting and structural steel. The operating cost (OPEX) savings is the savings in furnace fuel but this must be offset against the value of the lost VHP steam production (which notionally must be made up in a utility boiler elsewhere) and the additional electricity for the two fans.

2) It can be seen that the payback period is three years based on European gas prices and five years based on American gas prices.

Naphtha Furnace using Feed preheat

Naphtha is normally stored at ambient temperature and is preheated to 60°C using quench water before it is routed to the furnaces. Most cracking furnaces are equipped with a Feed Preheater (FPH) bank in the convection section. If flue gas is utilized to preheat naphtha from ambient conditions the stack temperature can be lowered to the desired range of sixty to seventy degrees Celsius. Till now designers and operators have not taken full advantage of this mainly because of the higher CAPEX and fear of dew point corrosion.

One of the consequences of this feature will be that the quench water duty previously absorbed by the naphtha will now shift to the air coolers in the quench water circuit. The additional electrical power that the air cooler fans will require will be about 10 KW per furnace.

Please refer to Table 3 for an overview of the performance of a naphtha furnace with and without an extended feed preheater. Please note that the Base Case naphtha furnace is the same as the one considered in the APH option.

1) The overall efficiency of the furnace can be improved from 94.4 to 96.0% for this option.

2) Unlike the APH option a combustion air fan will **not** be required. The furnace will continue to operate in induced draft mode.

3) The fuel firing and the VHP steam generation will drop marginally.

4) The upper tube rows of the FPH will be below the water dew point of the flue gas and some condensation will occur. As a precaution we have considered that the top four rows will be of stainless steel.

5) The flue gas NOx concentration will be unaffected by this option. However the total hourly NOx emissions and load on the DeNOx unit, if any, will decrease to the same extent as the fuel firing, which is around 0.6%.

Table 4 summarizes the results of the economic analysis of this option.

1) The capital cost (CAPEX) of this option covers the incremental cost of the extended FPH, extra ducting and structural steel. The OPEX savings is the reduction in furnace fuel after adjusting for the reduction in VHP steam production and the additional electricity for the ID fan and the Quench Water Air Cooler fans.

2) In the Base Case the naphtha feed piping between the external feed preheater and the convection section is insulated for heat conservation. In the FPH option naphtha will be fed at ambient temperature to the furnace and so these lines do not have to be insulated. This will yield some capital cost saving but this has not been taken into account in our analysis.

3) The payback period excluding CO2 credit will be three and a half years with European natural gas and five and a half years with US natural gas.

Ethane Furnace using Air Preheat

Fig 2 shows the Base Case Ethane furnace that was used in the study. The APH arrangement on an ethane furnace will be quite similar to the one on the naphtha furnace. Table 5 shows the impact that the APH would have on the furnace performance. The main points are as follows:

1) The overall efficiency of the furnace will improve from 91.7 to 96.5% for the APH option. This is mainly due to the lower stack temperature but also partly due to the smaller excess air.

2) In this furnace also the APH will require a forced draft combustion air fan. This will allow the excess air to be reduced from 8% to 5% which will result in lower stack losses and a better efficiency.

3) The fuel firing will go down by 11.0%, from 8.2 to 7.3 t/h. The absorbed heat will drop from 116.8 to 110.6 MW. As a result of these the VHP steam generation will go down from 97.0 to 88.8 t/h.

4) A portion of the APH will operate below the flue gas water dew point. Consequently the unit must be designed to prevent water accumulation and corrosion.

5) The impact on NOx was evaluated here also. The rise in air temperature will have a greater impact on NOx than the fall in excess air. We estimate that as a result the flue gas NOx concentration will increase slightly (around two percent). This can be a cause for concern if the furnace does not have a DeNOx unit and is operating close to the limits imposed by authorities. However since the total firing will go down, the hourly NOx production will drop. Plants can consider discussing with authorities to obtain a higher allowable NOx limit,

   If the furnace has a DeNOx unit, the chemical load on the unit will reduce, which should help to obtain a longer catalyst life.

The operating cost savings and payback for this option have been summarized in Table 6.

1) The capital cost includes the cost of the APH, the combustion air fan, extra ducting and structural steel. The operating cost savings is the savings in fuel firing with an adjustment for the value of the reduced VHP steam generation and the extra electricity for the ID fan and FD fan.

2) The payback period with natural gas priced at European rates will be one and a half years. The corresponding figure for US gas prices will be three years.

Ethane Furnace using Feed Preheat

Using feed at ambient temperature does not significantly reduce the stack temperature in a gas furnace. This is because the feed preheat curve has a steeper slope than the flue gas cooling curve, causing a pinch on the hot side of the Feed Preheater. Fig 4 shows the heating and cooling curves of ethane, boiler feed water and flue gas in the convection section. The hatched shape on the plot shows the maximum additional heat duty that can be recovered into feed, over and above the Base Case. The plot clearly shows that when the ethane feed temperature is lowered from the base value of 55°C to 30°C, the stack temperature will drop from 170°C to about 150°C. The flue gas cannot be cooled any more than this modest amount.

For this reason the feed preheat option was not investigated any further.

Conclusions and Summary

The key results from the different study options are listed together below.

Cracking furnace feed	Naphtha		Naphtha		Ethane	
Cold sink	Combustion air		Hydrocarbon feed		Combustion air	
	EU nat. gas	US nat. gas	EU nat. gas	US nat. gas	EU nat. gas	US nat. gas
OPEX savings	€ 0.60 mill	€ 0.36 mill	€ 0.17 mill	€ 0.10 mill	€ 1.27 mill	€ 0.77 mill
Payback time	3 years	5 years	3.5 years	5.5 years	1.5 years	3 years
Salient features of each option	Small number of additional equipment		No additional equipment		Small number of additional equipment	
	Higher CAPEX		Lower CAPEX		Higher CAPEX	
	More OPEX savings		Less OPEX savings		Best OPEX savings	
	Slight effect on NOx concentration		No effect on NOx		Around 2% higher NOx concentration	

All three scenarios have attractive payback periods and will be worth considering in high energy cost countries such as Europe and Asia. In US and other energy rich nations, the economic incentive will be much lower. In the US, where ethane cracking is common and increasing even more, improvement of furnace efficiency with air preheat can be worth considering as the payback time is less than 3 years.

If the options are compared against each other, in our opinion the Air preheating option has more potential, irrespective of feedstock.

The Feed Preheating option in liquid furnaces is cheaper but it also gives proportionally modest savings. Moreover, as earlier mentioned, the feed preheat duty is being shifted from one waste heat source (quench water) to another (flue gas). This option will not lead to a net improvement or cash savings in the plant as a whole unless there is a shortage of quench water and there are other, better uses for the saved quench water duty

Acknowledgements

The authors wish to express their thanks to their colleagues Mr Ashok Laghate, Ms Susan Cooper, Mr Sachin Singare and others in Technip Benelux who contributed to this study.

References

1) API 560: Fired Heaters for General Refinery Service, 4th edition

2) API 535: Burners for Fired Heaters in General Refinery Service, 2nd edition

Table 1

Naphtha Furnace with Combustion Air Preheating – Performance Summary

		Base Case	Air Preheat Case
Naphtha feed	t/h	60.8	60.8
Ambient conditions	-	15°C, 85% R.H.	15°C, 85% R.H.
Air temp. to burners	°C	15	77
Stack temperature	°C	108	61
Excess air	%	8	5
Absorbed heat	MW	113.2	110.4
Overall efficiency	%	94.4	97.0
Fuel flow	t/h	8.4	8.0
Fired heat	MW	119.9	113.9
VHP steam production	t/h	79.66	76.01
NOx concentration		Base	Close to Base

Table 2

Economics of Naphtha Furnace with Combustion Air Preheating

		Air Preheat Case (EU location)	Air Preheat Case (US location)
Savings in fuel firing	million €/yr	1.42	0.86
Cost of lost VHP steam capacity	million €/yr	0.73	0.44
Cost of additional electricity	million €/yr	0.10	0.06
Net OPEX saving	million €/yr	0.60	0.36
Payback	Years	3	5

Table 3
Naphtha Furnace with Feed Preheating – Performance Summary

		Base Case	Feed Preheat Case
Naphtha feed	t/h	60.8	60.8
Feed temperature	-	60°C	30°C
Combustion air temp.	°C	15	15
Stack temperature	°C	108	71
Excess air	%	8	8
Absorbed heat	MW	113.2	114.3
Overall efficiency	%	94.4	96.0
Fuel flow	t/h	8.4	8.35
Fired heat	MW	119.9	119.1
VHP steam production	t/h	79.66	79.63
NOx concentration		Base	Base

Table 4
Economics of Naphtha Furnace with Feed Preheating

		Feed Preheat Case (EU location)	Feed Preheat Case (US location)
Savings in fuel firing	million €/yr	0.19	0.12
Cost of lost VHP steam capacity	million €/yr	< 0.01	< 0.01
Cost of additional electricity	million €/yr	0.01	< 0.01
Net OPEX saving	million €/yr	0.17	0.10
Payback	Years	3.5	5.5

Table 5
Ethane Furnace with Air Preheating – Performance Summary

		Base Case	Air Preheat Case
Ethane feed	t/h	51.9	51.9
Ambient conditions	-	15°C, 85% R.H.	15°C, 85% R.H.
Air temp. to burners	°C	15	129
Stack temperature	°C	169	64
Excess air	%	10	5
Absorbed heat	MW	116.8	110.6
Overall efficiency	%	91.7	96.5
Fuel flow	t/h	8.2	7.3
Fired heat	MW	127.3	114.5
VHP steam production	t/h	97.0	88.75
NOx concentration		Base	Base + 2%

Table 6
Economics of Ethane Furnace with Air Preheating

		Air Preheat Case (EU location)	Air Preheat Case (US location)
Savings in fuel firing	million €/yr	3.01	1.83
Cost of lost VHP steam capacity	million €/yr	1,64	1.00
Cost of additional electricity	million €/yr	0.10	0.06
Net OPEX saving	million €/yr	1.27	0.77
Payback	Years	1.5	3

Fig 1
Naphtha cracking furnace

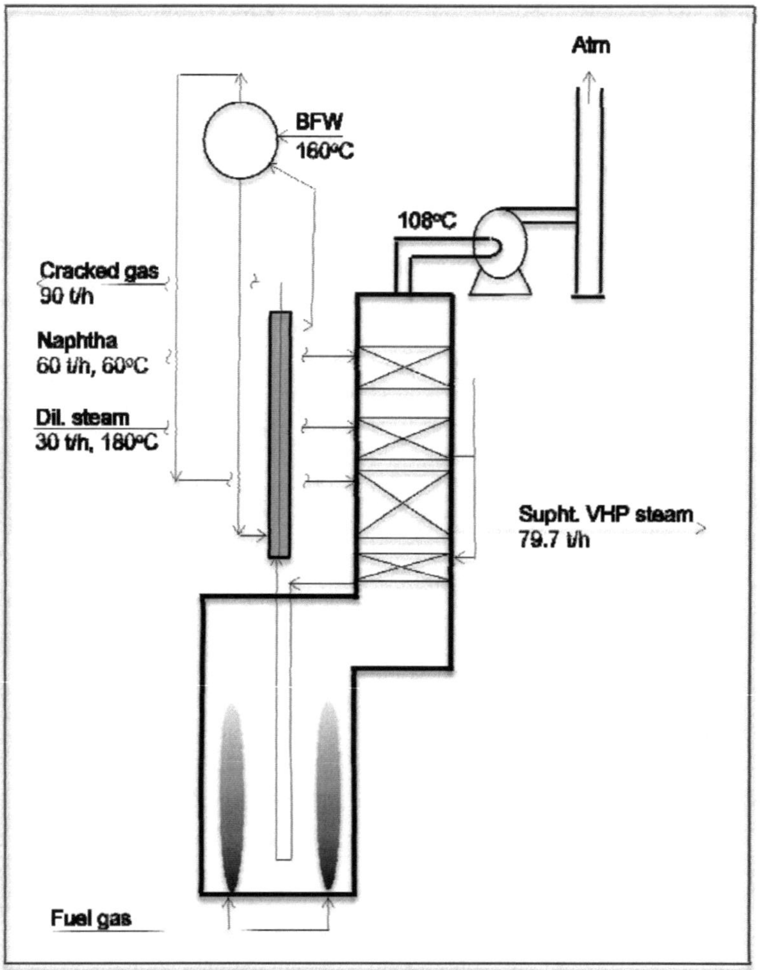

Fig 2
Ethane cracking furnace

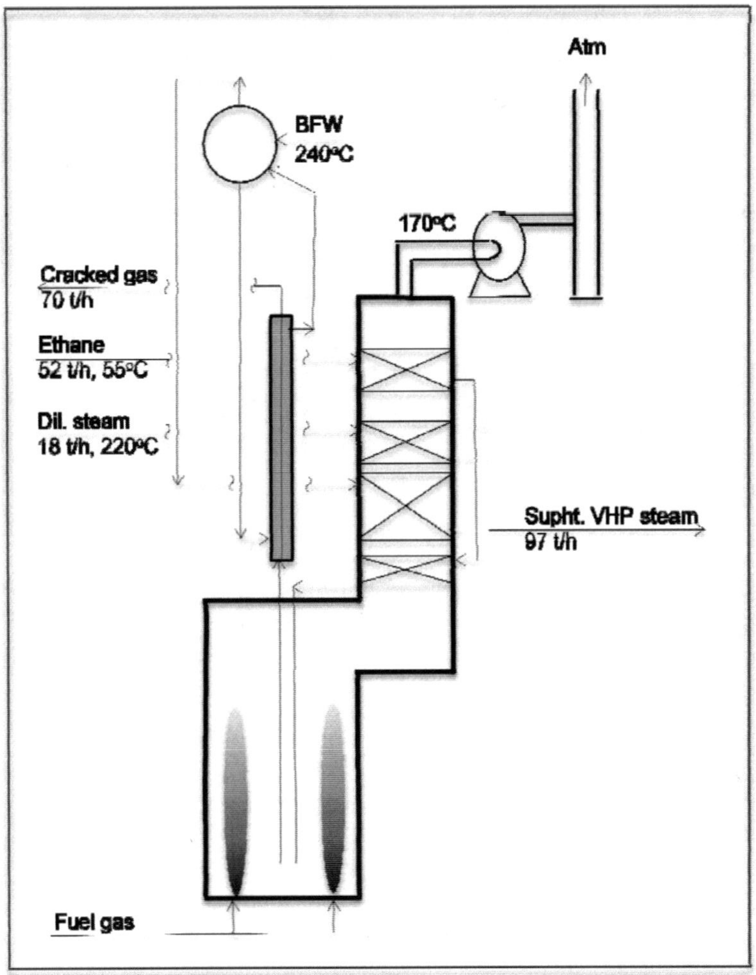

Fig 3

Naphtha cracking furnace – Air Preheater option

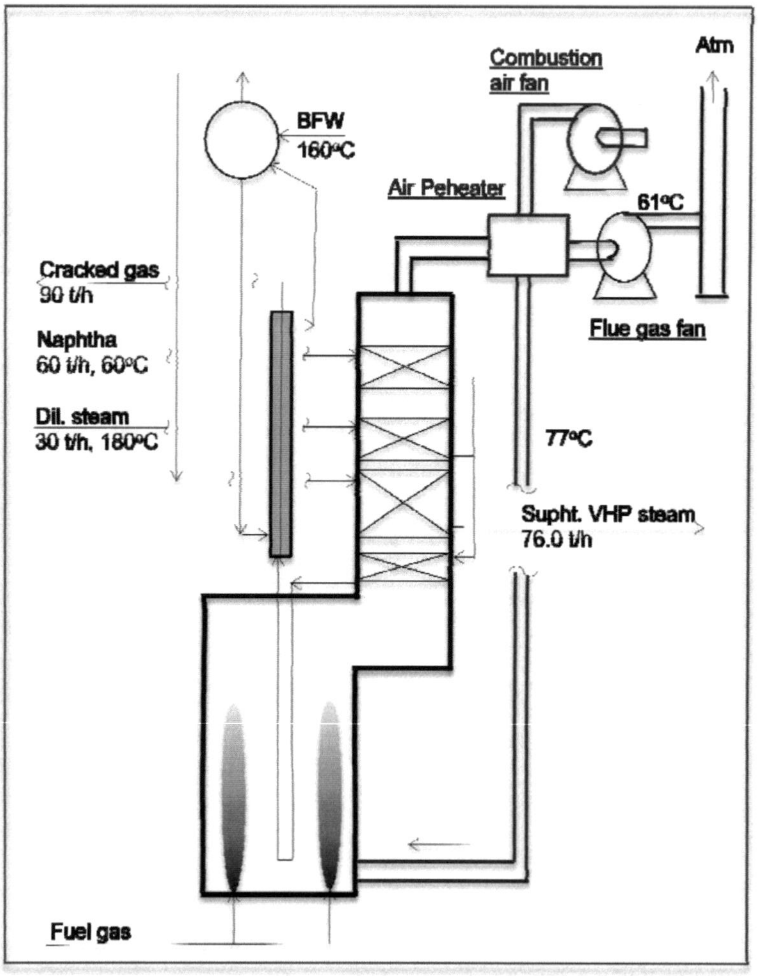

Fig 4

<u>Ethane cracking furnace – Feed Preheater option</u>

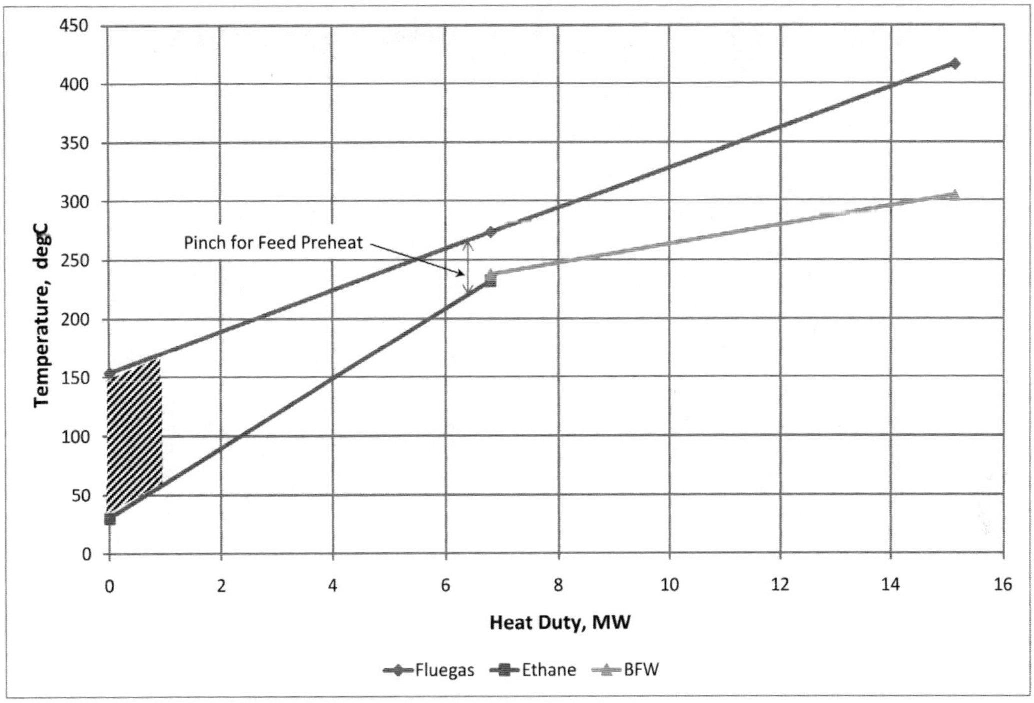

Modernize Your Control Functionality
Easy Steps to Start Process Optimization

13 AIChE
Spring Meeting, San Antonio, TX

Presenters

- Laurie Ben
- James Beall

EMERSON.
Process Management

Agenda

- Introduction
- Examples – Best Practices & Lessons Learned
  - Split Range Duty
  - Smith Predictor
  - Gain Scheduling/Adaptive Control
  - Signal Characterization
- Summary

Introduction

- Modernizing Control System – Decision Done Check!
- 'Do I' or 'Don't I' take advantage of new technology
  - Implemented older cumbersome techniques?
  - Complex custom strategies that handle automation challenges?
  - Hard to implement & maintain?
- Standard features that are field proven, practical, and easy to implement
  - Easier to implement & maintain
  - More efficient and effective control strategies
  - Achieve improvements in process performance
  - Lays the foundation for optimization

Agenda

- Introduction
- Examples – Best Practices & Lessons Learned
  - Split Range Duty
    - Control Challenge
    - Legacy Solution
    - Newer Technology Solution
    - Summary
  - Smith Predictor
  - Gain Scheduling/Adaptive Control
  - Signal Characterization
- Summary

Split Range Duty – Control Challenge

- Use steam and hot oil as energy source for distillation column for temperature controller.
- Hot oil is "free energy" and should be maximized
- Control Scenarios
  - Maximize hot oil and supplement with steam
  - Base load steam (if meter bad) and supplement with hot oil
  - Base load hot oil flow (if bad meter) and supplement with steam
- Constraints
  - Minimum and maximum hot oil valve position and flow rate
  - Minimum steam flow rate

Split Range Duty – Legacy Solution

- Old legacy system implemented this feature using a velocity PID with incremental output and a "micro expert rules system"

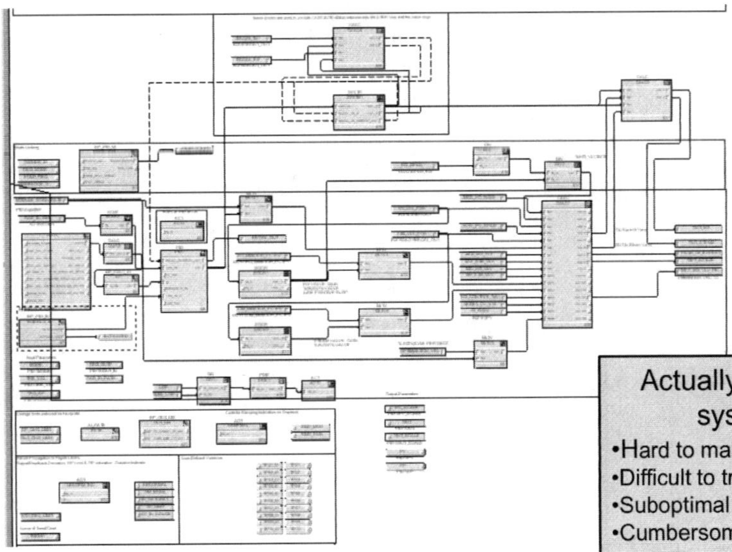

Actually Duplicated in New system, but still…

- Hard to maintain and modify
- Difficult to troubleshoot
- Suboptimal performance
- Cumbersome/Kludge/Ugly

Split Range Duty – New Technology Solution

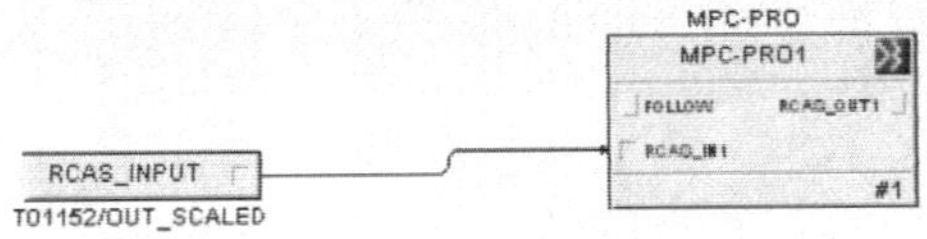

Embedded MPC Function Block provides all control functionality! EASY!!!

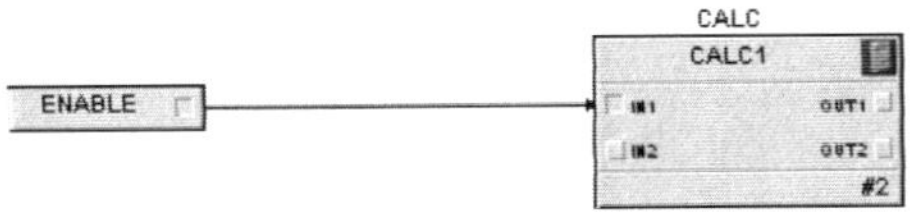

Simple calculations to calculate limits

Split Range Duty – New Technology Solution

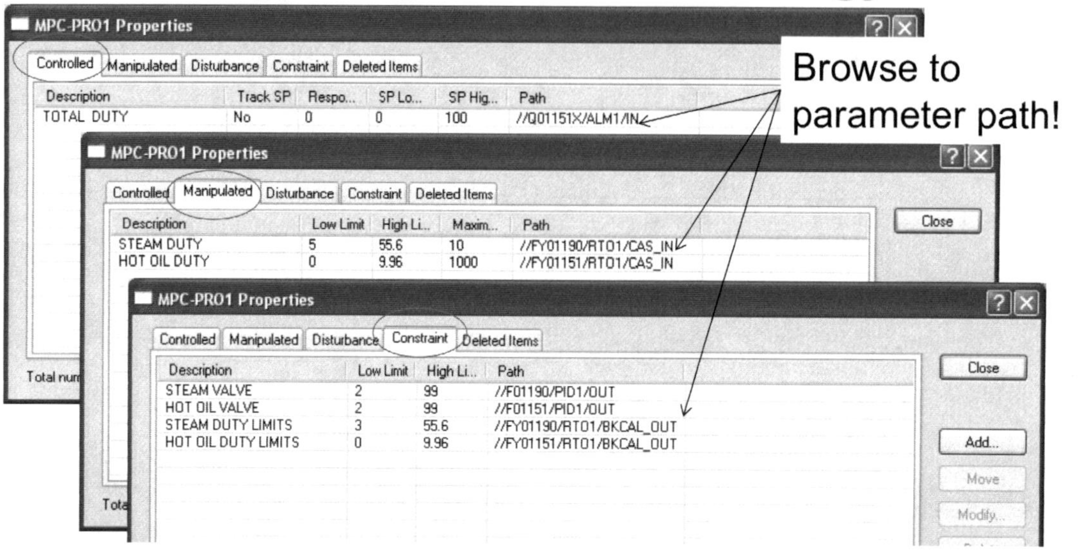

Split Range Duty – New Technology Solution

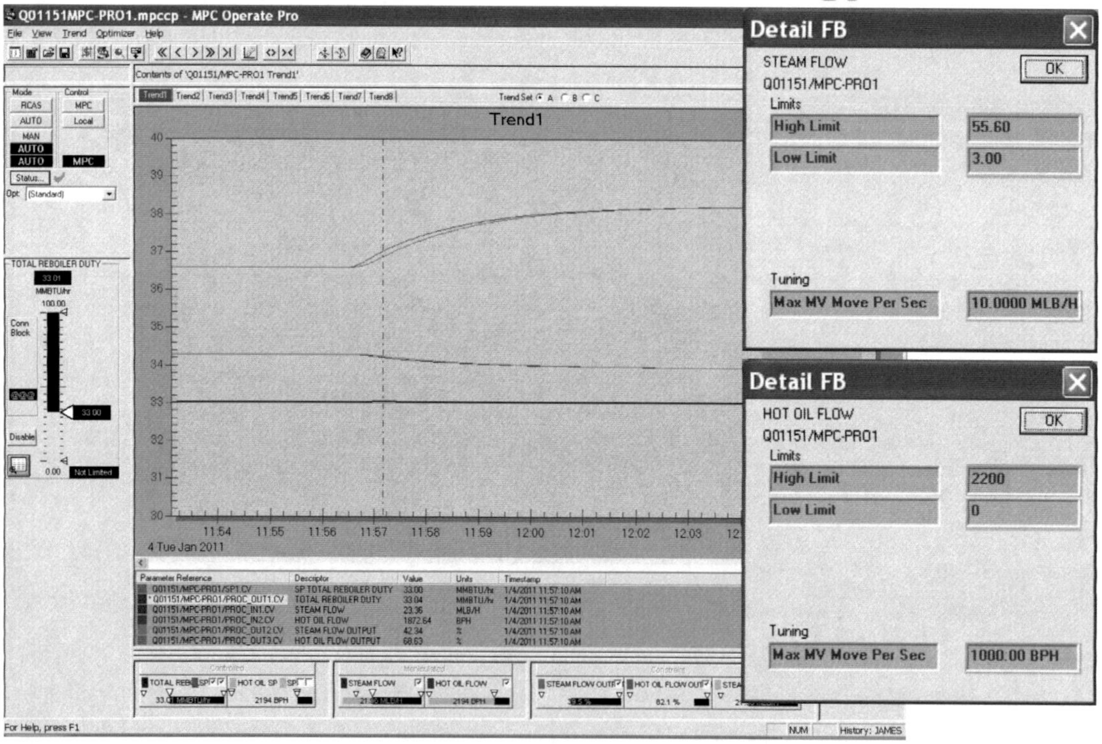

Summary Split Range Duty Control

Embedded MPC (New Technology) vs Legacy Solution

- Better performing solution
  - Bumpless transfers and constraint control
- Easier to Maintain
  - Change limits and settings
  - Simpler implementation
  - Easy to understand, train others, document
- Easily expands to handle future constraints
- Reduce energy costs!

Agenda

- Introduction
- Examples – Best Practices & Lessons Learned
  - Split Range Duty
  - Smith Predictor
    - Control Challenge
    - Legacy Solution
    - Newer Technology Solution
    - Summary
  - Gain Scheduling/Adaptive Control
  - Signal Characterization
- Summary

Smith-Predictor PID – Control Challenge

- Control of processes that are <u>"dead time dominant"</u> (dead time > time constant)
- Typically <u>custom configured</u>
  - some legacy control systems have preconfigured options, e.g. PROVOX
- Algorithm uses a process model
  - (gain, dead time and time constant)
- <u>Challenge</u>: as process dynamics change from model used, performance degrades rapidly

Smith-Predictor PID – Legacy Solution

- Traditional Smith-Predictor

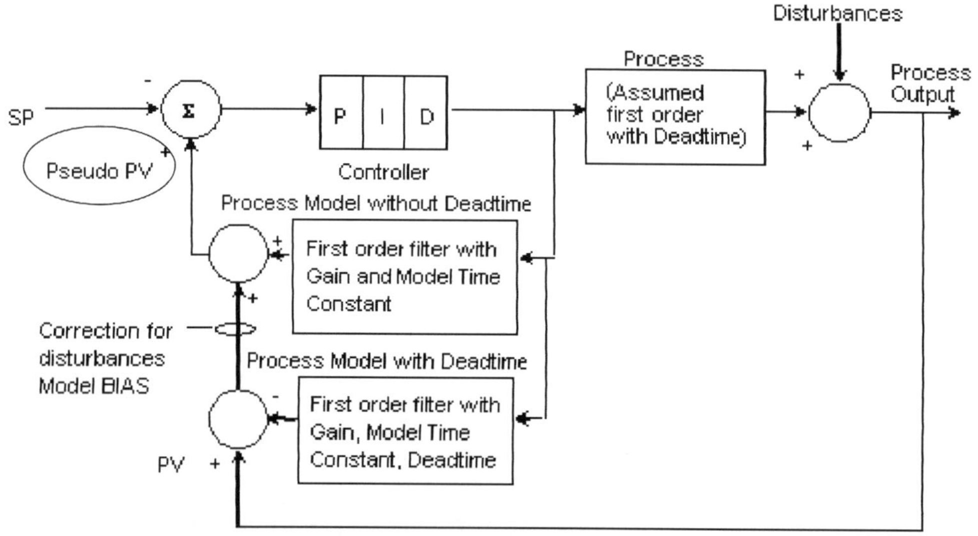

•High maintenance when process dynamics change
•Performance degrades rapidly as process dynamics change from configured model

Smith-Predictor PID – New Technology Solution #1

Easy to implement the Smith-Predictor but…

- High Maintenance when process dynamics change
- Performance degrades rapidly as process dynamics change from configured model

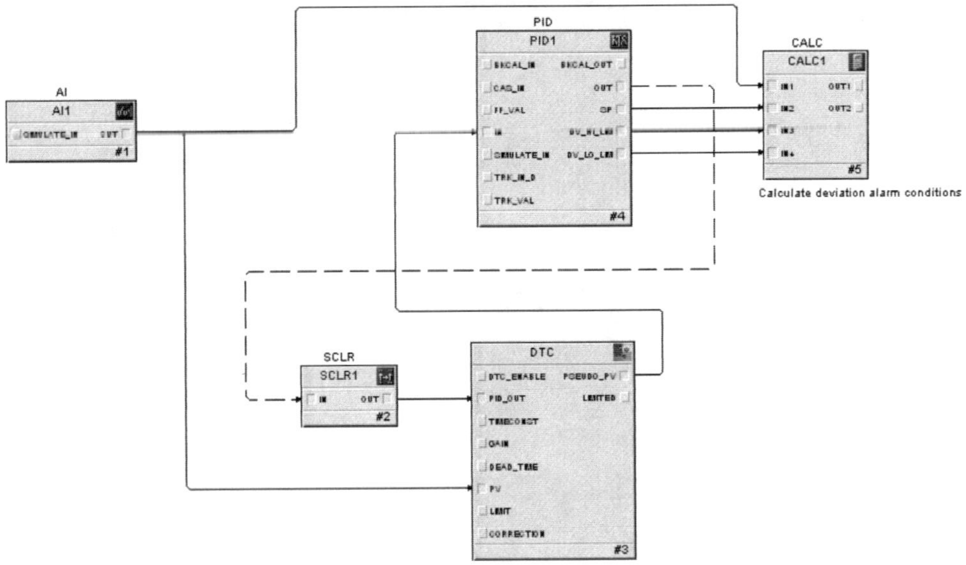

Smith-Predictor PID – New Technology Solution #2

- Embedded MPC Solution

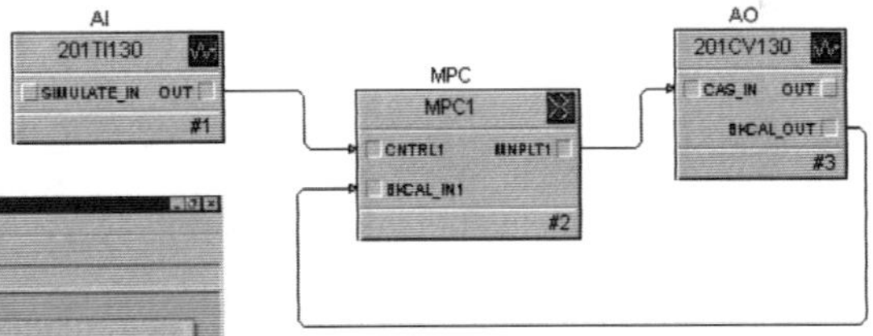

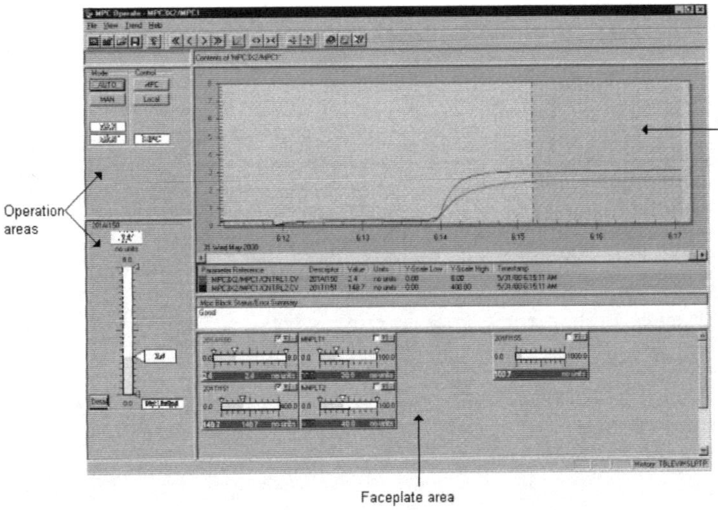

- Provides Future Prediction
- Provides MPC Faceplate – no configuration needed
- More robust under changing process dynamics

Smith-Predictor PID – New Technology Solution #2

- Embedded MPC Solution

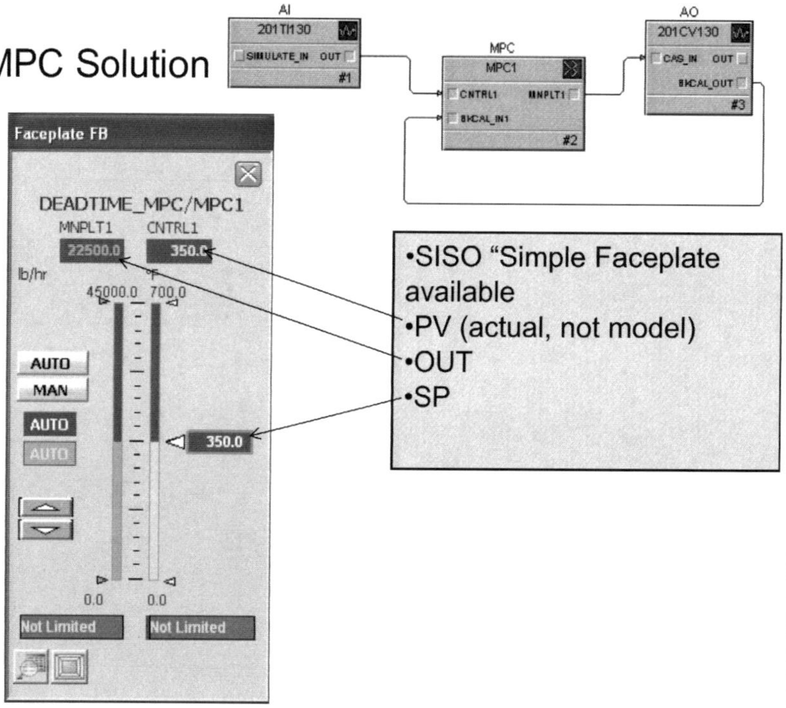

Summary Smith-Predictor PID

New Technology Solutions

- Standard Smith Predictor (#1) and Embedded MPC (#2)
- Both easy to implement and understand!

New Technology Solution #2 (Embedded MPC)

- More robust solution under a changing process model
- Provides visual display of the "future" PV
- Save money through better control performance

Agenda

- Introduction
- Examples – Best Practices & Lessons Learned
  - Split Range Duty
  - Smith Predictor
  - Gain Scheduling/Adaptive Control
    - Control Challenge
    - Legacy Solution
    - New Technology Solution
    - Summary
  - Signal Characterization
- Summary

Gain Scheduling – Control Challenge

- Legacy systems have limited capability to modify tuning based on process conditions
- Others had add-on application software, e.g. PROVOX "Gain Scheduler" system provided 3 sets of Gain, Reset, and Rate applied in 3 separate regions of a "state variable"
  - State variable = PV, OUT or "other"
  - Transition between tuning sets interpolated over "dead band"
- Beware: some suspend loop execution when tuning parameters change, so don't change the tuning every scan

Gain Scheduling – Legacy Solution

- Custom code
- Multiple points, registers, displays
- Restricted (slower) update rate

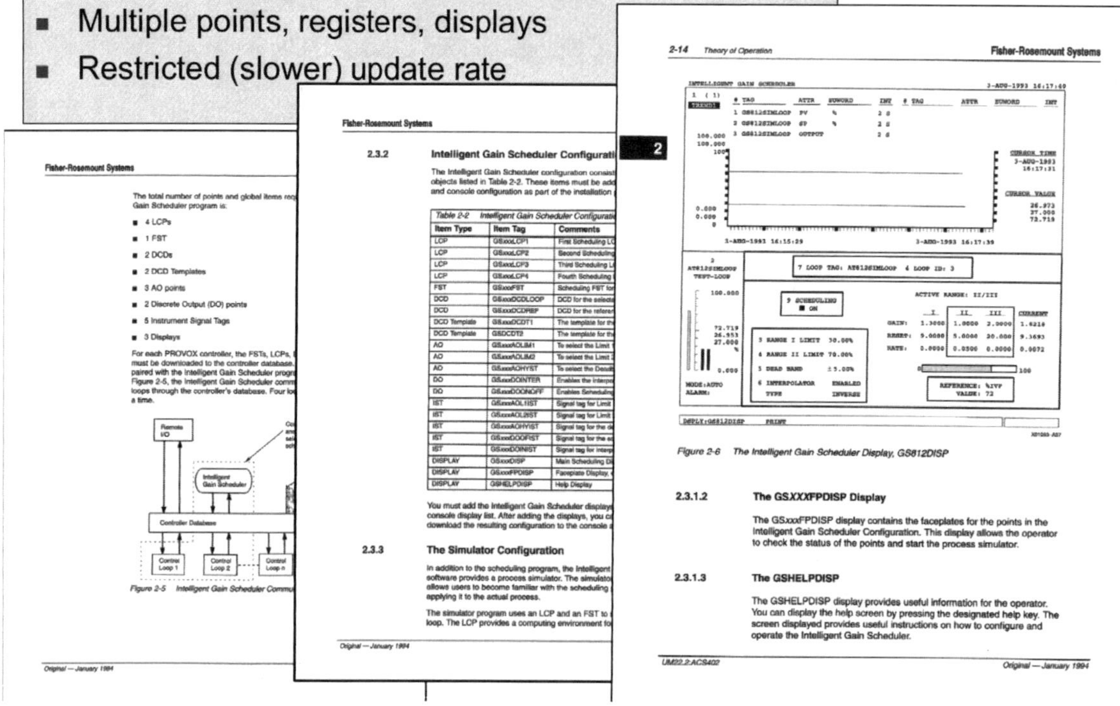

Gain Scheduling – New Technology Solution #1

Standard Modules

- Standard configuration
- Preconfigured faceplate & detailed display
- Easy to implement
- Tuning updates at loop execution

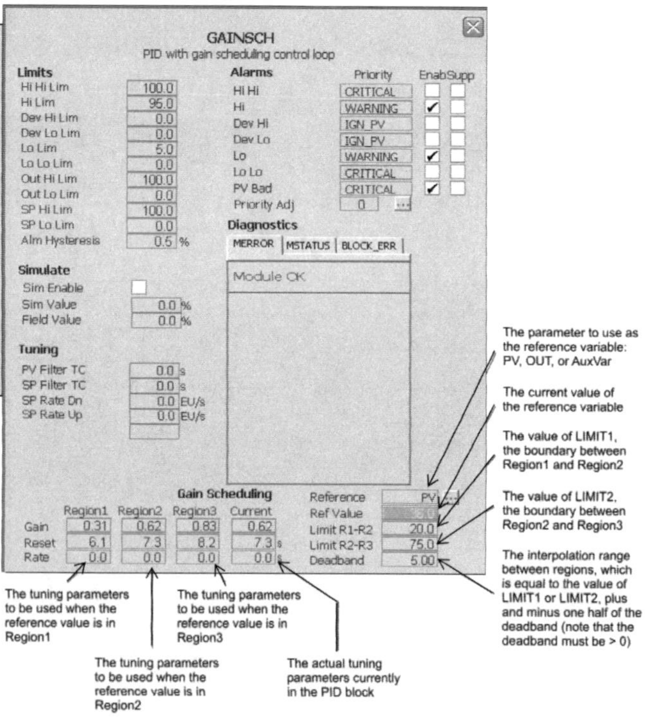

Gain Scheduling – New Technology Solution #2

New Technology "Adaptive Control"

- "Learns" process models from normal operating moves
- Allows up to 5 tuning regions of the "state variable"
- Partial Adaptive Control
  - Use "approved model" only & selected tuning rule in each region
  - Lambda PI or IMC PID; "Closed loop time constant" or "Lambda Factor".
- Optional Full Adaptive Control
  - continuous "Learning" of process model in each region
  - updates tuning while in each region per selected tuning rule
  - set allowable deviation (%) from approved model in each region

New Technology – Adaptive Control

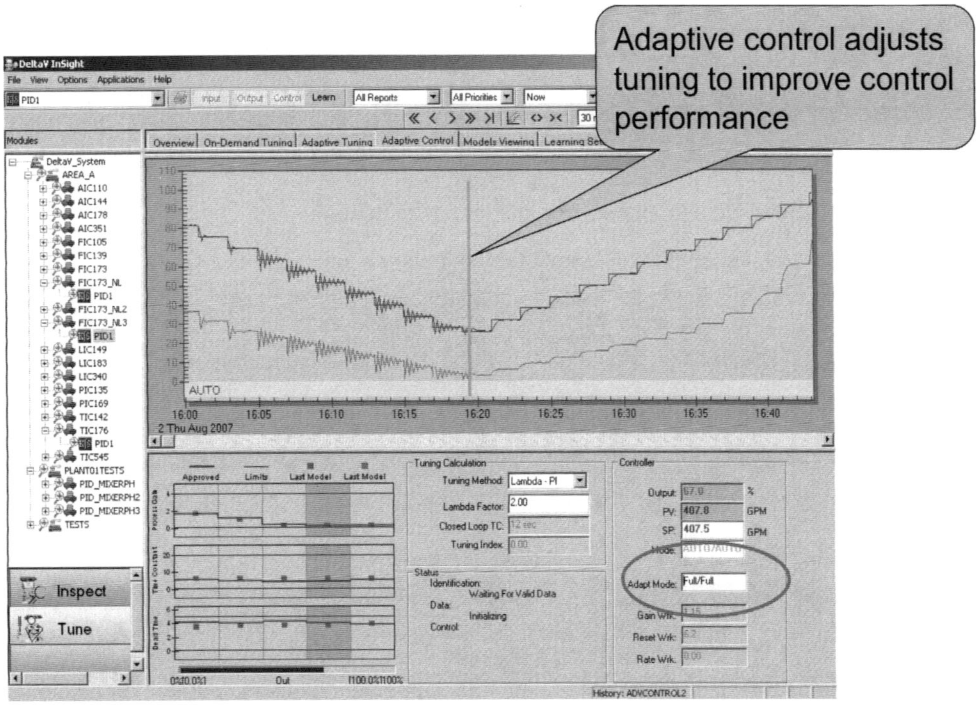

Summary Gain Scheduling

New Technology Solution(s) vs. Legacy Solution

- Gain Scheduling easy to implement and understand!
  - Standard configuration
  - Preconfigured faceplate & detailed display
  - Tuning updates at loop execution
- Adaptive Control provides additional benefit and ease of implementation
  - Learns models during normal operation – NO STEP TESTs!
  - More Regions – Up to 5
  - Automatic tuning updates when process model changes
- Improved control performance saves money!

Summary Gain Scheduling

New Technology Solution(s) vs. Legacy Solution

- Gain Scheduling easy to implement and understand!
  - Standard configuration
  - Preconfigured faceplate & detailed display
  - Tuning updates at loop execution
- Adaptive Control provides additional benefit and ease of implementation
  - Learns models during normal operation – NO STEP TESTs!
  - More Regions – Up to 5
  - Automatic tuning updates when process model changes
- Improved control performance saves money!

Agenda

- Introduction
- Examples – Best Practices & Lessons Learned
  - Split Range Duty
  - Smith Predictor
  - Gain Scheduling/Adaptive Control
  - Signal Characterization
    - Control Challenge
    - Legacy Solution
    - New Technology Solution
    - Summary
- Summary

Signal Characterization - Control Challenge

- Input Characterization (AI)
  - Combustion control – characterize fuel and air
  - Level – characterize pv (e.g. level to volume)
- Output characterization (AO)
  - modify installed flow characteristics of valves, dampers or other final control elements
  - Use to modify loop gain, e.g. output path to slave loop
- Legacy System Challenges
  - Lack of resolution (few points)
  - External black box characterizers may shift, lose documentation
  - Potential operator/maintenance confusion on signal going to final control element

Signal Characterization – Legacy Solution

- Legacy system using external analog characterizer

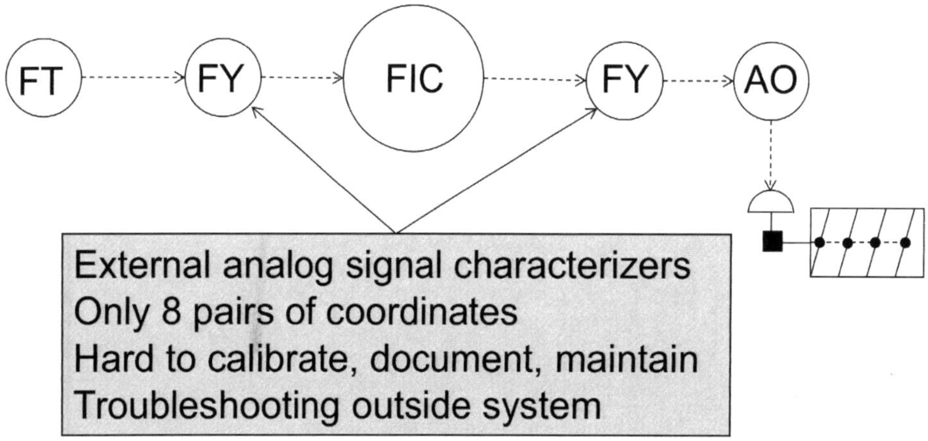

Signal Characterization – New Technology Solution #1

- Standard Module implementation provides 21 point signal characterizer function block (SGCR)

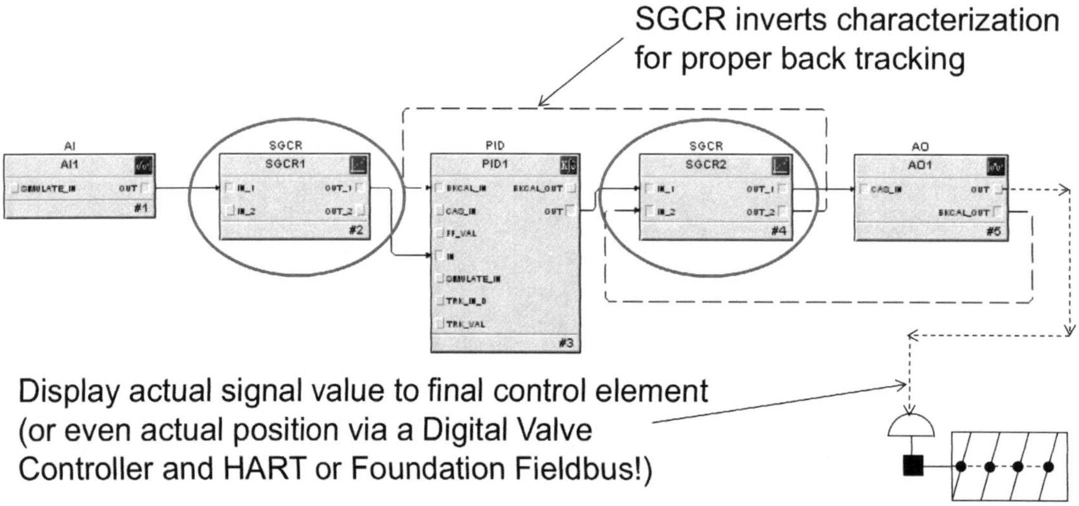

Signal Characterization – New Technology Solution #2

- Adaptive Control
  - Adjust tuning to compensate for final control element characteristics

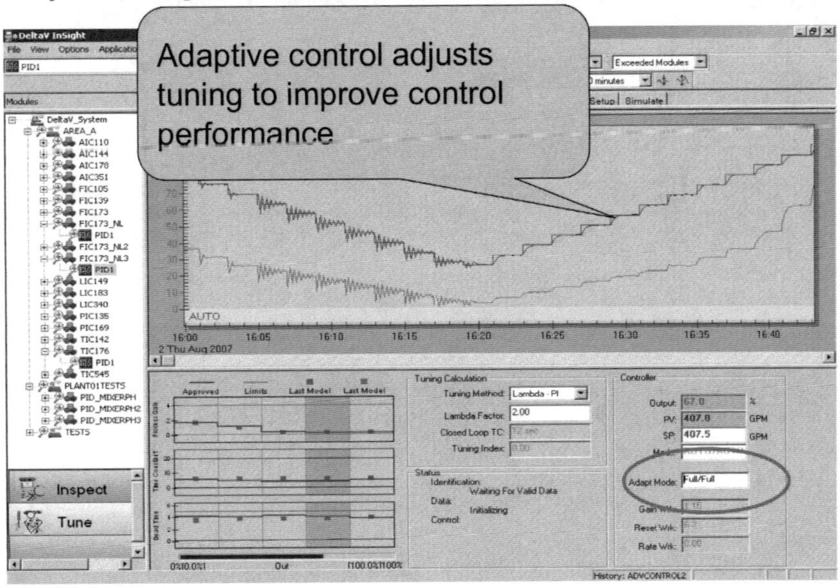

Summary Signal Characterization

- Newer Technology Signal Characterization function block
  - Better resolution (21 pairs of coordinates)
  - Self documenting (exists in control system)
  - No calibration required vs. 'external black box'
  - Provides "inverse" characterization for analog output characterizations
  - Reduce confusion: display signal being sent to final control element
- Applying Adaptive Tuning may provide a superior solution for AO characterization of final control element

Agenda

- Introduction
- Examples – Best Practices & Lessons Learned
  - Split Range Duty
  - Smith Predictor
  - Gain Scheduling/Adaptive Control
  - Controller Characterization
- Summary

Summary

- Practical application of control features that deal with common challenges
- Highlighted best practices & lessons learned
  - Split Range Duty
  - Smith Predictor
  - Gain Scheduling/Adaptive Control
  - Signal Characterization
- Do look at new functionality available when modernizing
  - Allows jump start to process optimization
  - Leads to improved process performance
  - Easier to implement and maintain over the life investment

Contacts & Questions?

James.Beall@Emerson.com
903-235-7935

Laurie.Ben@Emerson.com
512-834-7237

Paralleled Double-Effect Distillation: Simulative Case Studies

*Pei Wang, Leilei Dai, Ming Shi , Fan Wang, Hong Gao, Jinsheng Sun**
School of Chemical Engineering and Technology, Tianjin University, No. 92 Wei Jin Road, Tianjin 300072, P. R. China

Abstract

The increasing scarcity and growing demand for energy have led to an increasingly attention to the energy-efficient methods. Double-effect distillation is a common method to achieve remarkable energy conservation, which has three kinds of typical configurations including the feed splitting (FS), the light split/forward (LSF) and the light split/reverse (LSR). In spite of a good pathway for energy saving, better options are expected to cut down energy expenditure by a bigger margin than double-effect distillation does.

In this paper, case studies of paralleled or combined double-effect distillation schemes are proposed on the basis of current double-effect distillation. In order to verify the feasibility of this method, the improved methanol distillation scheme and organosilicon distillation scheme are optimized under this principle as examples. Exergy losses and the energy consumptions are measured as a tool to compare the two prevalent schemes with their corresponding optimized counterparts, respectively. The results of the total exergy loss and the energy consumption of these schemes show that the optimized schemes can cut the corner of the exergy loss by 22.8% and 40.41% and the energy consumption by 15.61% and 44.28%, respectively, in comparison to their popular cousins.

Keywords: double-effect distillation, paralleled double-effect distillation, methanol distillation, organosilicon monomer distillation, distillation process simulation, energy saving, exergy analysis

1. Introduction

Distillation, in chemical process and related industries, is by far the most widely applied for the separation of liquid mixture. In spite of its well-known benefit and widespread application, the major drawback is its high energy consumption. In classical distillation process, heat is transferred to the reboiler of distillation column. Along with the distillation column to separate the mixture, a large number of heat are directly discharge into the surrounding environment and cooling water. Araujo AB [1] has done some research to achieve that distillation has a relatively low thermodynamic efficiency and it requires the input of high quality energy in the reboiler to push the separation of liquid mixture. Therefore, the focus of energy conservation is to consider how to recover or reuse the discharged heat, or how to reduce the quantity of transferred heat to the column. In the early 20th century, many scholars have devoted to the study of saving energy in distillation, and now most of correlative techniques are widely used in distillation process [2,3]. These energy saving methods involve multi-effect distillation, heat pump distillation, heat integrated distillation and etc [3-8]. But without extra rotating machines by electricity or other power producers, multi-effect distillation has been more favored in research and application [9].

Many researches confirmed multi-effect distillation to be one of the effective methods to reduce the energy consumption [10-12]. The energy saving effect of those schemes used multi-effect distillation is associated with the number of the effects and the characteristics of separation system. Generally speaking, those schemes can offer energy savings around 30% with respect to the original or classical multicomponent distillation scheme [23, 27, 28]. Considering the interaction between investment of plants and energy saving effect, the double-effect distillation has been the most commonly used in not only research but also field application.

The use of the double-effect distillation technology for optimizing the conventional distillation schemes can cut a large amount of the energy requirement in comparison with the correspondingly prevalent schemes [13]. Although the system is very difficult to control in the sense that it is nonlinear, multivariable and interacting [14]. More and more published studies [15-19] have been focused on further questing the way of energy saving on the basis of the double-effect distillation schemes. However, the overhead vapor of one high pressured column only can integrate with the reboiler of one low pressured column in those published studies. In view of this, in this paper, the paralleled or combined double-effect distillation is proposed to broaden the thought for the application of the double-effect distillation, in which the overhead vapor from one of the high pressured column is divided into two or more streams to successively integrate with the reboiler of the other columns [9].

2. *Theory*

2.1 Double-effect or heat integrated distillation system

A double-effect distillation system is made up of two adjacent distillation columns running in different pressure, which can lead to a temperature difference between the top of the high pressured column and the bottom of the low pressured column. Through the thermodynamic connection, the reboiler of the low pressure column plays a role of the condenser for the high pressure column. According to the direction of the mass flow and the heat integration, the double-effect distillation can fall into three kinds of configuration: the feed splitting (FS), the light split/forward (LSF) and the light split/reverse (LSR), as showed in Figure. 1.

One of the configurations with a high thermodynamic efficiency is the feed splitting (FS), in which the feed is divided roughly equal between the two columns running in different pressure, and the products can be obtained from both of the top and bottom of the two columns, as illustrated in figure. 1 (a). The overhead vapor form the high pressured column is used to heat the reboiler from the low pressured column. There is only energy exchange between the two columns, so the heat duties of the feed splitting (FS) are approximately equal to the heat duty for single column with the same conditions divided by the number of effect.

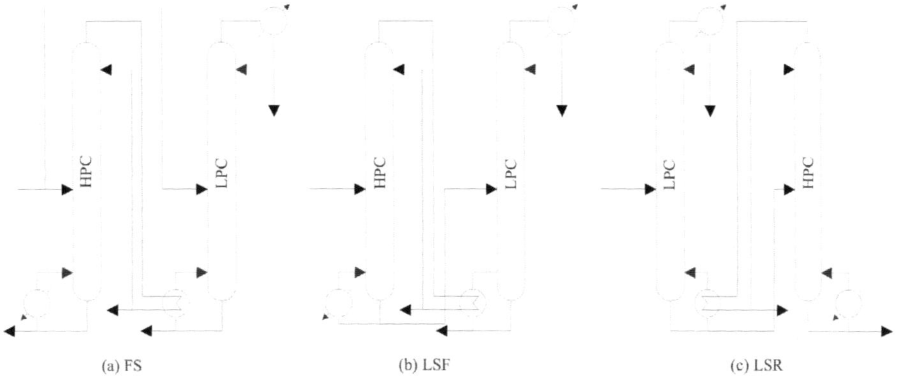

Fig. 1. Three configurations of double-effect distillation

Figure. 1 (b) and (c) respectively shows the configuration of the light split/forward heat integration (LSF) and the light split/reverse heat integration (LSR). Different from the feed splitting (FS), the total feed stream was sent to one column in the other two configurations. In the light split/forward (LSF), the total feed is pumped to the high-pressure column (HPC) and the bottom stream of HPC flows into the low-pressure column (LPC). In this way, all of the heavy components are from LPC and the light components from the top both of the two columns. Figure. 1 (b) indicates the direction of mass flow and heat integration is the same. The last configuration is the light split/reverse (LSR), as shown in figure. 1 (c), all of the feed is sent to the low-pressure column (LPC). It can remove the two kinds of light components from the top both of the two columns and one kind of heavy components from the bottom of HPC. Moreover, the direction of heat integration is in the reverse with the mass flow. These two kinds of configurations have not only energy exchange but also substantial exchange, therefore, when the operating parameters of one column are changed, the other column will also change.

Theoretically, the three kinds of configurations are empirically 30% less energy consumption than the single column distillation. As for choosing what kind of the configurations of the double-effect distillation, it depends on the properties of the multi-component mixture, the pressure of saturated steam, the temperature of cooling water, structure parameters of existing distillation column and so on.

2.2 Exergy analysis of the distillation process

According to the First Law of Thermodynamics, also called the Law of Energetic Balances, no matter what happens, the total amount of energy has not yet changed. However, when the various types of energy generate the mutual conversion, the quality of the energy is always decreased. The separation process in distillation is irreversible. In order to separate the mixture in distillation column, it needs the driving force which must consume a certain amount of external work, by this means, the decline of the quality of energy is inevitable. Therefore, we should choose a reasonable driving force in distillation process to avoid unnecessary loss of the quality of energy. It defined the concept of exergy (E_x) to compare the quality of the different types of energy. Exergy, which is a measure for evaluating the extent of available energy or the quality of energy, is based on the Second Law of Thermodynamics. Exergy analysis reveals the essence of the energy, and can be used as an important tool to evaluate the effect of energy saving. And it usually combines with process optimization studies, and as its guidance, to identify cost saving opportunities and amend the existing scheme.

Szargut [29] has summarized a comprehensive definition of the exergy, that is: 'Exergy is the amount of work obtainable when some matter is brought to a state thermodynamic equilibrium with the common components of the natural surroundings by means of reversible processes, involving interaction only with the above mentioned components of nature'. The state of one stream recorded as 1, according to the definition of exergy, the exergy of the stream can be expressed as:

$$E_x = (H_1 - H_0) - T_0(S_1 - S_0) \qquad (2)$$

where, H_1 and S_1 is the enthalpy and the entropy of the stream 1, and T_0 is the temperature of the surrounding and H_0 and S_0 represent the enthalpy and the entropy of the surrounding.

The state of another stream recorded as 2, when the system stream change from state 1 (T_1, P_1) to state 2 (T_2, P_2), the variation of exergy (ΔE_x) is given by

$$\Delta E_x = E_{x,1} - E_{x,2} = \Delta H - T_0 \Delta S \qquad (3)$$

In the conventional distillation process, it must input the heat or energy to drive the occurrence of separation process. The bottom of the column input the heat (Q_R) at this temperature (T_R), and the top of the column was sent to the heat (Q_C) at this temperature (T_C). The net work consumption (W) in distillation is

$$W = Q_R(1 - \frac{T_0}{T_R}) - Q_C(1 - \frac{T_0}{T_C}) \quad (4)$$

When the enthalpy difference between the product and the feed is very small, $Q_R=Q_C=Q$ can be approximately gotten, then Equation (4) is rewritten to be:

$$W = QT_0(\frac{1}{T_C} - \frac{1}{T_R}) \quad (5)$$

Then, the total exergy loss (Ex_{loss}) of the distillation column can be calculated by

$$Ex_{Loss} = W - \Delta E_x \quad (6)$$

3. *Prototype case: the prevalent distillation schemes*

3.1 The improved four-column methanol distillation scheme

In the methanol industry, it has gradually appeared two-column, three-column, four-column and five-column distillation scheme [20-23]. With the increase of the number of distillation column, the corresponding distillation scheme can not only increase the efficiency of the process, but also cut the energy consumption. At present, the four-column scheme has a dominant status to produce the purified methanol. However, the improved four-column distillation scheme can cut some more energy requirement with respect to the traditional four-column distillation scheme. The improved four-column distillation scheme is seen as the comparative reference in the paper.

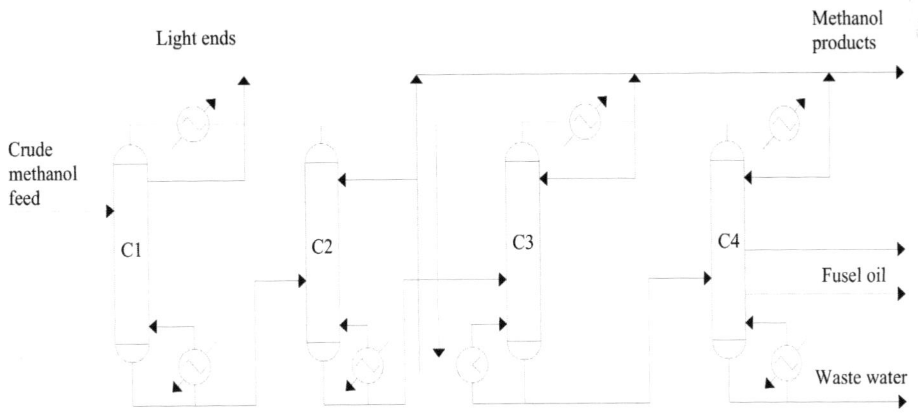

Fig. 2. The improved four-column double-effect methanol distillation scheme (C1, light ends column; C2, pressured column; C3, atmospheric column; C4, water column)

In the improved four-column distillation scheme, as shown in figure. 2 [24], the crude methanol flows into the light ends column (C1) in 87.5 ℃ for removing the light ends and the remnants of dissolved gas from the top of C1. Besides, in order to raise the methanol recovery rates, the light ends column (C1) designs a two-stage condenser system. The crude methanol was distilled successively from the pressured column (C2) and the atmospheric column (C3), and most of the methanol products can be received at the top both of C2 and C3. The bottom stream of C1, C2 and C3 is pumped respectively as the feed into C2, C3 and the water column (C4). Residual methanol, fusel oil and waste water are extracted from the corresponding position of the top, side stream and the bottom of the water column (C4).

In the improved scheme, the operation pressure of C2 is 0.9MPa, and the other three columns run under the atmospheric pressure. Due to the different of operating pressure, the overhead vapor of C2 can undertake thermal coupling connection with the reboiler of C3.

3.2 The current organosilicon monomer distillation scheme

According to the survey, there are several kinds of organosilicon distillation schemes. However, the 9-column organosilicon distillation scheme [9], in China, has an important role in the organosilicon industrial, as illustrated in Figure. 3, it is used as the base case.

Figure. 3 shows that the crude monomer is delivered to the heavies removal column (C1') and the bottom stream of C1' has a lot of the unsteady substances of high boiling points. These unsteady substances of high boiling points are pumped to the high boiling point column (C9') as the feed, and the products from the top of C9' continue to serve as the crude monomer by circumfluence. By this way, it can accomplish the separation between the unsteady substances of high boiling points and the useful substances at a relatively low temperature in the high boiling point column (C9') and avoid the unsteady substances of high boiling points from thermal cracking to generate many unknown substances which would greatly affect the subsequent separation of product. The low removal column (C2') provides trimethylchlorosilane (Me3), methyldichlorosilane (MeH), and other light components for lights removal column (C3'), MeH column (C4'), azeotrope removal column (C5') and Me3 column (C6'), which can finally achieve two products (MeH and Me3) and three wastes. It also provides methyltrichlorosilane (Me1) and dimethylchlorosilane (Me2) and other heavy components for Me1 column (C7') and Me2 column (C8'), which can finally get two products (Me1 and Me2).

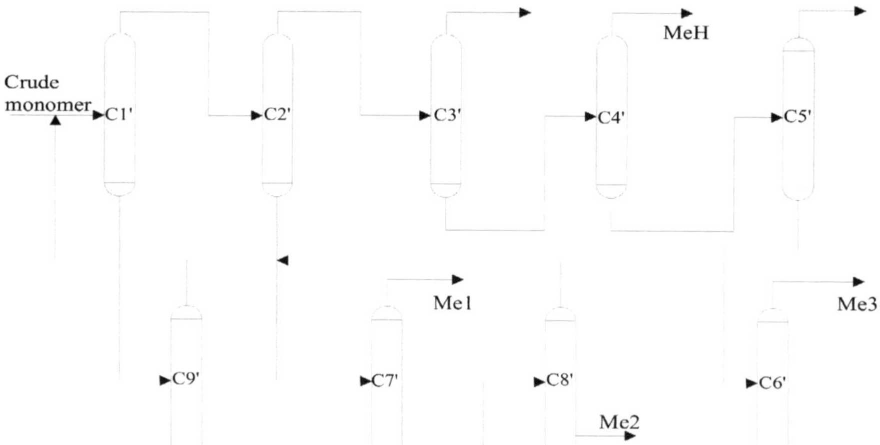

Fig. 3. The prevailing 9-column distillation scheme (C1', heavies removal column; C2', low removal column; C3', lights removal column; C4', MeH column; C5', azeotrope removal column; C6', Me3 column; C7', Me1 column; C8', Me2 column; C9', high boiling point column)

In this scheme, the operating pressure of all nine columns is relatively low, and the highest operating pressure is 260 kPa in the lights removal column (C3'). Although it has a different operating pressure between C3' and the other columns, but the temperature difference is not enough to make them use double-effect or heat integrated distillation.

4. Simulation and analysis both of the prevalent schemes and the optimized schemes

4.1 The novel five-column heat integrated methanol distillation scheme

Although the energy saving of the improved four-column methanol distillation scheme can reach 10% compared with the traditional four-column scheme, it still exist several shortages, which will be illustrated by the results of computer simulation to the improved four-column methanol distillation scheme. The flow rate of the crude methanol is 239230 kg/hr, and the temperature is 87.5 ℃, and the composition of the crude methanol is exhibited in table 1. The inlet and outlet temperature of the cooling water are 30 ℃ and 40 ℃.The methanol content is specified as 99.99 wt% in the methanol products and 10ppm in the wastewater of the water column. Table 2 shows the operating parameters of each column in the improved four-column scheme.

Table 1. Crude methanol feed data

Component	Mass fraction%	Component	Mass fraction%
H_2	0.0100	Pentanol	0.0338
N_2	0.0710	Formic acid	0.0030
Methane	0.0100	Acetic acid	0.0030
CO	0.0360	Methyl formate	0.0092
CO_2	0.6640	Acetone	0.0193
Methanol	95.0000	Methyl ethyl ketone	0.0030
Dimethyl ether	0.0309	Diisopropylether	0.0056
Trimethylamine	0.0010	Water	3.7100
Ethanol	0.0900	i-Pentane	0.0562
i-Propanol	0.0024	n-Pentane	0.0169
Propanol	0.0264	n-Hexane	0.0487
i-Butanol	0.0338	n-Heptane	0.0009
Butanol	0.0185	n-Octane	0.0050
3-Pentanol	0.1407	n-Nonane	0.0050

Table 2. The operating parameters of improved four-column scheme

The operating parameters	C1	C2	C3	C4
Top pressure (kPa)	150	900	110	110
Top temperature (℃)	81.0	132.6	67.7	67.7
Bottom temperature (℃)	83.8	139.1	96.5	104.5
Reflux ratio	0.58	2.44	1.68	11.57
Number of theoretical stage	33	52	58	61

The results of simulation indicate that the hot utility is still in a high level because the reboilers of C1, C2 and (C4) still need the saturated steam to transmit the heat. Table 2 shows the reflux ratio of C4 is 11.57 and this large reflux ratio can cause the increasing of the energy requirement of the water column (C4). Besides, according to the flowsheet of this scheme, the diameter of column is so large that the emergence of many urgent cases.

In view of above these defects in the improved four-column scheme, the novel five-column heat integrated methanol distillation scheme (shown as Fig. 4) is designed in this paper to reduce the impact of these factors as far as possible by using the paralleled or combined double-effect distillation. Juntao Zhang [23] and Jinsheng Sun [25] are also proposed a five-column methanol distillation scheme by using triple-effect distillation and double-effect distillation respectively, and the effect of energy saving of the two schemes is outstanding.

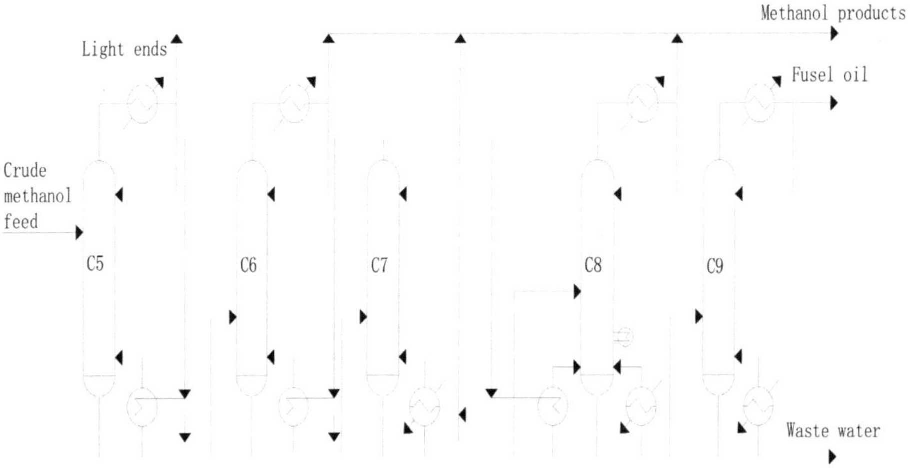

Fig. 4. The five-column heat integrated methanol distillation scheme (C5, light ends column; C6, prepositive atmospheric column; C7, pressured column; C8, postpositive atmospheric column; C9, water column)

According to the principle of the paralleled double-effect distillation, the four columns are unable to constitute the heat integration network in the improved four-column scheme. Through repeated verification, the prepositive atmospheric column (C6) is added after the light ends column (C5), meanwhile, the operating pressure of the pressured column (C7) is reduced to increase its throughput. In this way, the overhead vapor of C7 is split into three streams to respectively integrate with the light ends column (C5), the prepositive atmospheric column (C6) and the postpositive atmospheric column (C8). Due to use the paralleled double-effect distillation, the duties of the hot utility have decreased in a great margin. However, there are still part of heat from the background process is absorbed by the cooling water. So an inter-reboiler with a low temperature level is designed in the postpositive atmospheric column (C8) and it can share the heat with the background process streams, besides, we have proved the feasibility by using the column grand composite curve analysis of columns. Figure. 4 shows that the product of the top of the water column (C9) is fusel oil instead of purified methanol, which can not only reduce the reflux ratio of C9, but also make the diameter of column smaller to reduce the difficulty of operation.

Due to produce the fusel oil instead of purified methanol in the water column (C9) in the novel five-column distillation scheme, the methanol content of the feed for the water column have to limit at 0.016 wt%. In the five-column scheme, the operating pressure and the number of theoretical stage for each column are determined through the existing process, the onsite data and the requirement of the heat coupling. The proportion of the methanol products from the C6, C7 and C8 is determined by their own double-effect or heat integrated distillation system. The simulation results of the novel five-column scheme are listed in table 3.

Table 3. The operating parameters of five-column scheme

Column	C5	C6	C7	C8	C9
Top pressure (kPa)	160	110	400	110	110
Top temperature (℃)	83	66.6	104	66.6	86.5
Bottom temperature (℃)	85.6	84	118.8	104.5	103.2
Reflux ratio	1.46	1.49	2.23	1.87	4
Number of theoretical stage	33	54	54	65	29

4.2 Energy consumption and exergy loss analyses for the two methanol schemes

In the improved four-column scheme, the pressured column and the atmospheric column constitute the distillation system of heat coupling, there are three columns (C1, C2, C4) need the hot utility, and the entire duty of reboiler is 145.4×10^3 kw. As described above, the light ends column (C5) and the prepositive atmospheric column (C6) don't need the saturated steam in the five-column scheme and the total duty of reboiler is reduced to 122.7×10^3 kw. The detailed data of the reboiler duty as shown in table 4.

Table 4. The duty of reboiler in the two schemes

Column	C1	C2	C4	C7	C8	C9
Duty of reboiler (10^4 kw)	3.30	10.04	1.20	8.97	2.86	0.44

Obviously, the novel five-column scheme is much better than the improved four-column scheme in energy saving and the energy consumption of the novel scheme can be reduced by 15.61%. However, the recovery rate of methanol of the novel scheme is approximately equal with the improved four-column scheme.

Only the energy consumption analysis is not revealed the energy saving essence of the novel five-column scheme. Exergy analyses are used to further compare the two schemes, and the exergy flow rates are calculated through using the computer simulation data. The activity coefficient of the liquid is calculated using the Alcohol Package thermodynamic model. The calculation of the net work consumption of distillation column should pay more attention to the temperature of the cooling water. When the cooling water exchanges heat with other streams, the temperature of the cooling water is momentarily changed, so it is replaced the temperature of cooling water with the arithmetic average of inlet and outlet temperature of cooling water. Figure. 5 show the overall exergy loss of each column in the two methanol distillation schemes, in addition to the overall exergy loss of C6. Due to optimize the four-column scheme, the prepositive atmospheric column (C6) is added in the transformed scheme, and the overall exergy loss of C6 is 9157.6 kw. The exergy loss of each column, besides the pressured column (C7), in the novel five-column scheme is obviously less than their counterparts in the improved four-column scheme.

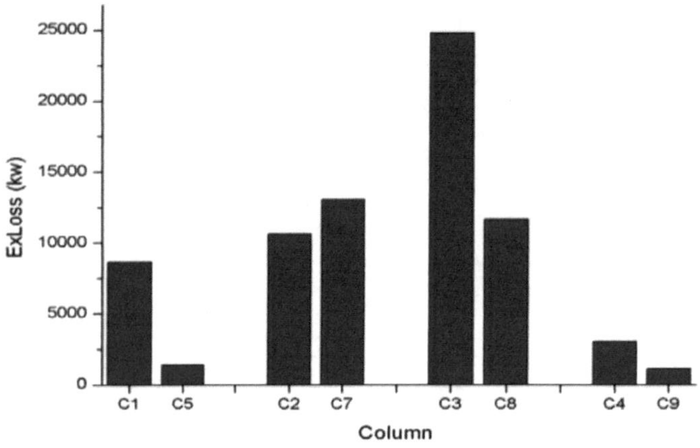

Fig. 5. The overall exergy loss analysis of each column in the two schemes

The total exergy loss of the improved four-column scheme and the transformed scheme is, respectively, 47100.1 kw and 36355.1 kw. The total exergy loss of the optimized scheme is dropped by 22.8%. Through the comparison between the exergy loss and energy consumption of the two schemes, it is indicated that the novel five-column scheme should be applied to produce the purified methanol in industry.

4.3 The novel heat coupling organosilicon distillation scheme

The primary source of the crude organosilicon monomers is the catalytic reaction products between chloromethane and silicon at high temperature and the table 5 displays the normal boiling points and mass percentage of the main components of the crude organosilicon monomers. Methyltrichlorosilane (Mel), dimethylchlorosilane (Me2), trimethylchlorosilane (Me3) and methyldichlorosilane (MeH) are the most valuable components in the crude organosilicon monomers. The purification requirements on organosilicon of these materials are shown in table 6.

Table 5. The crude organosilicon monomer feed data

Component	Normal boiling point at 1 atm/℃	Mass percentage %
LB		1
Methyldichlorosilane	41.8	4
Silicon tetrachloride	57.6	0.5
Trimethylchlorosilane	57.9	2.5
Methyltrichlorosilane	66.4	11
Dimethyldichlorosilane	70.2	80
HB		1
Temperature (℃)	40	
Pressure（kPa）	100	
Flow rate (kg/h)	26200	

Table 6. Purification demand on organosilicon

Monomer	Purification demand
MeH	99.5%
Me3	99.5%
Me1	99.5%
Me2	99.95%

The conventional nine-column scheme is simulated with the aid of existing process and onsite data to achieve the operating parameters of each column, as illustrated in table 7. The bottom temperature of the heavies removal column (C1') and the high boiling point column (C9') are obviously much bigger than that of the other seven columns in table 7. According to the principle of the double-effect distillation, the top temperature of the high pressured column is relatively high in comparison with the bottom temperature of the low pressured column, and the high pressured column can integrate with the low pressured column. Although the top temperature of C1' is less than the bottom temperature of the other seven columns, the overhead vapor of C1' still has a great opportunity to conduct heat with the other seven columns with the increasing of its top temperature.

Table 7. The simulation results of prevailing 9-column

Column	C1'	C2'	C3'	C4'	C5'	C6'	C7'	C8'	C9'
Top temperature (℃)	70.8	65.3	57.3	66.7	72.1	70.6	71.2	74.9	72.7
Bottom temperature (℃)	124.1	89.1	80.4	85.2	74.9	75.4	75.3	75.3	126.5
Reboiler duty (MW)	11.67	5.61	1.13	0.51	1.19	0.46	12.27	16.08	0.02
Condenser duty (MW)	-12.47	-5.53	-1.08	-0.53	-1.20	-0.47	-12.69	-16.12	-0.02
Top pressure (kPa)	110	180	260	220	170	150	110	110	110

According to the way of the paralleled double-effect distillation, the increasing operating pressure in the heavies removal column (C1') can make the top temperature of C1' ascending to integrate with the other seven columns. The idea is still in study to overcome some difficulties. However, we have got another feasible method, in which the operating pressure of C1' and the Me1 column (C7') is raised to 360 kPa and 280 kPa, respectively. In this way, the top temperature of C1' and C7' is increased to ensure they can integrate with the other columns, besides the high boiling point column, and the parallel double-effect or heat integrated organosilicon monomer distillation scheme is designed as shown in figure. 6 [26]. The feasibility of this heat coupling scheme has confirmed through three x-y diagrams including the heavies component and Me1, the heavies component and Me2, Me1 and Me2 [9].

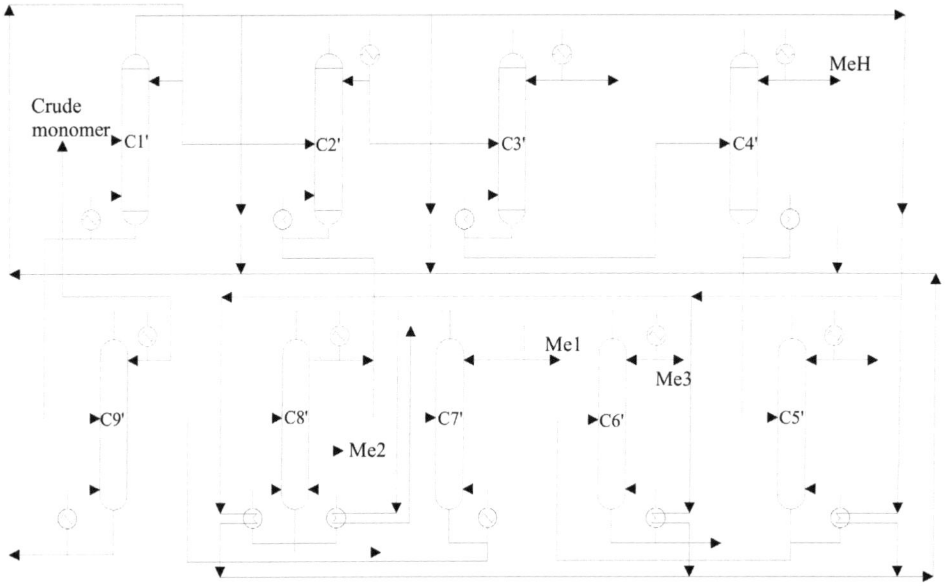

Fig. 6. The heat coupling 9-column scheme (C1', heavies removal column; C2', low removal column; C3', lights removal column; C4', MeH column; C5', azeotrope removal column; C6', Me3 column; C7', Me1 column; C8', Me2 column; C9', high boiling point column)

Figure. 6 shows the heavies removal column (C1') make the heat coupling with the low removal column (C2'), the lights removal column (C3'), the MeH column (C4'), the azeotrope removal column (C5'), the Me3 column (C6') and the Me2 column (C8'), successively. In order to gain a better energy saving effect, C8' is designed with two reboilers, the one integrated with C1' and the other integrated with the Me1 column (C7'). The operating parameters and the duty of reboiler and condenser of each column in the heat integrated 9-column are enumerated in table 8.

Table 8. The simulation results of heat integrated 9-column

Column	C1'	C2'	C3'	C4'	C5'	C6'	C7'	C8'	C9'
Top temperature (℃)	114.3	65.3	57.3	66.7	72.1	70.6	105.1	74.9	72.7
Bottom temperature (℃)	124.9	89.9	78.8	84.5	75.4	75.4	110.9	76.6	121.4
Reboiler duty (MW)	13.06	5.62	1.13	0.51	1.19	0.46	13.83	14.12	0.38
Condenser duty (MW)	-13.72	-5.54	-1.08	-0.53	-1.20	-0.47	-14.02	-14.43	-0.41
Top pressure (kPa)	360	180	260	220	170	150	280	110	110

4.4 Energy consumption and exergy loss analyses for the two organosilicon schemes

Figure. 4 indicate that each column of the prevalent nine-column scheme is not integrated with the other columns and the reboilers of all of these columns are required the saturated steam (600 kPa). According to the data from the table 7, the total duty of reboiler is 48.94 MW for the prevailing 9-column scheme. In the new nine-column scheme, C1' integrate with C2', C3', C4', C5', C6' and C8' successively and C7' integrate with C8', so C2', C3', C4', C5', C6' and C8' don't consume the saturated steam. The total duty of reboiler is 27.27 MW. Obviously, the heat integrated 9-column scheme has an absolute advantage over the prevailing 9-column scheme in energy saving, and the energy savings can reach 44.28%.

As the same with the methanol distillation scheme, the inlet and outlet temperature of cooling water are 30 ℃ and 40 ℃, and the total exergy loss is calculated through the results of computer simulation of the two schemes. The overall exergy loss comparison of each column in the two schemes is illustrated in figure. 7. In the normal scheme, the throughput of C9' is so small that the overall exergy loss of C9' is very near to zero, so it only have the exergy loss of C9' of the new scheme in figure. 7. The overall exergy loss of each column in the optimized scheme is declined in comparison with that of the corresponding column in the normal scheme except the high boiling point column (C9'). Because the crude organosilicon monomers contains some unstable components, in order to avoid these components happening thermal cracking in the heavies removal column (C1') in the novel scheme, more liquid from the bottom of C1' is pumped to C9' as the feed. That can cause the ascent of the throughput and the overall exergy loss of C9'.

The total exergy loss is 14035.5 kw and 8363.2 kw, correspondingly, in the conventional scheme and the novel scheme, and the exergy loss of the new scheme can be reduced by 40.41%. Through the above analysis and comparison, the heat integrated nine-column organosilicon distillation scheme should be a substitute for the prevalent nine-column scheme.

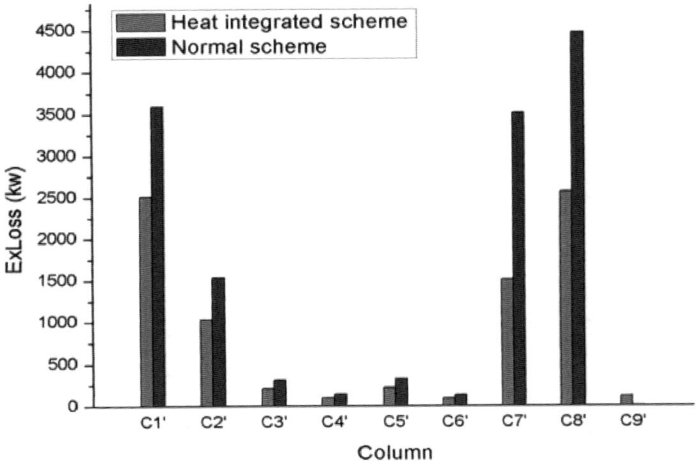

Fig. 7. The exergy loss comparison between the heat integrated scheme and the normal scheme

5. *Conclusions*

Paralleled double-effect or heat integrated distillation for energy saving is proposed, exampled by heat integrated five-column methanol distillation scheme and the heat coupling 9-column organosilicon monomer distillation scheme, with the aid of computer simulation. Besides, the optimized methanol distillation scheme and the transformed organosilicon scheme are comprehensively analyzed through the energy consumption and the exergy loss in comparison with the correspondingly prevalent scheme.

Adding a prepositive atmospheric column and decreasing the operating pressure of the pressured column in the novel five-column scheme can make the pressured column to simultaneously integrate with the light ends column, the prepositive atmospheric column and the postpositive atmospheric column. Besides, an inter-reboiler is added to integrate with the background process streams. The energy consumption and the exergy loss of the new scheme are 15.61% and 22.8% lower than the improved scheme.

Increasing the operating pressure of the heavies removal column and the Me1 column can make them to integrate with the other columns, in addition to the high boiling point column, to constitute a paralleled double-effect distillation system in the optimized nine-column scheme. The energy consumption and the exergy loss of the novel scheme are 44.28% and 40.41% less than the normal scheme.

The two cases expound that the key of paralleled double-effect distillation process lies in the reasonable adjustment of the energy distribution, the operating parameters and the configuration of some columns. It not only reduces the quantity of energy consumption, but also avoids the decrease of the quality of energy. For factories with a large numbers of distillation schemes of high energy requirement, the paralleled schemes lead to improve economic benefit.

References

[1] Araujo AB, Brito RP, Vasconcelos LS, "Exergetic analysis of distillation processes-a case study," Energy 2007;32:1185-93.

[2] Michele Mascia, Francesca Ferrara, Annalisa Vacca, Giuseppe Tola, Massimiliano Errico, "Design of heat integrated distillation systems for a light ends separation plant," Applied Thermal Engineering 2007;27:1205-1211.

[3] Amiya K. Jana, "Heat integrated distillation operation," Applied Energy 2010;87:1477-1494.

[4] Hilde K. Engelien, Sigurd Skogestad, "Multi-effect distillation applied to an industrial case study," Chemical Engineering and Processing 2005;44:819-826.

[5] James G. Gebbie, Michael K. Jensen, Piotr A. Domanski, "Experimental transient performance of a heat pump equipped with a distillation column," International Journal of Refrigeration 2007;30:499-505.

[6] Massimiliano Errico,Giuseppe Tola, Michele Mascia, "Energy saving in a crude distillation unit by a preflash implementation," Applied Thermal Engineering2009;29:1642-1647.

[7] Massimiliano Erricoa, Giuseppe Tola, Ben-Guang Rong, Daniele Demurtas, Ilkka Turunen, "Energy saving and capital cost evaluation in distillation column sequences with a divided wall column," chemical engineering research and design 2009;87:1649-1657.

[8] Ivar J. Halvorsen, Sigurd Skogestad, "Energy efficient distillation," Journal of Natural Gas Science and Engineering 2011;3:571-580.

[9] Jinsheng Sun, Fan Wang, Tingting Ma, Hong Gao, Yanzhen Liu, Fang Cai, "Exergy analysis of a parallel double-effect organosilicon monomer distillation scheme," Energy 2012;47:498-504.

[10] Hilde K. Engelien, Sigurd Skogestad, "Multi-effect distillation applied to an industrial case study," Chemical Engineering and Processing 2005;44:819-826.

[11] Kazuo Matsuda, Kenichi Kawazuishi, Yasuki Kansha, Chihiro Fushimi, Masaki Nagao, Hiroshi Kunikiyo, Fusao Masuda, Atsushi Tsutsumi, "Advanced energy saving in distillation process with self-heat recuperation technology," Energy 2011;36:4640-4645.

[12] M.A. Sharaf, A.S. Nafey, Lourdes García-Rodríguez, "Exergy and thermo-economic analyses of a combined solar organic cycle with multi-effect distillation (MED) desalination process," Desalination2011;272:135-147.

[13] Weiwen Wang, Wenjian Wang, "Energy Saving in Methylchlorosilane Distillation," International Conference on Computer Distributed Control and Intelligent Enviromental Monitoring 2012:290-293.

[14] Myungwan Han, Sunwon Park, "Multivariable control of double-effect distillation

configurations," J. Proc. Cont. 1996:247-253.
[15] Hilde K. Engelien, Sigurd Skogestad, "Multi-effect distillation applied to an industrial case study," Chemical Engineering and Processing2005;44:819-826.
[16] Debadrita Maiti, Amiya K. Jana, Amar Nath Samanta, "A novel heat integrated batch distillation scheme," Applied Energy2011;88:5221-5225.
[17] Anton A. Kiss, Servando J. Flores Landaeta, Carlos A. Infante Ferreira, "Towards energy efficient distillation technologies-Making the right choice," Energy2012;47:531-542.
[18] Diego C. Alarcón-Padilla, Lourdes García-Rodríguez, Julián Blanco-Gálvez, "Experimental assessment of connection of an absorption heat pump to a multi-effect distillation unit," Desalination2010;250:500-505.
[19] Diego C. Alarcón-Padilla, Lourdes García-Rodríguez, Julián Blanco-Gálvez, "Design recommendations for a multi-effect distillation plant connected to a double-effect absorption heat pump: A solar desalination case study," Desalination2010;262:11-14.
[20] Fiedler E, Grossman G. Methanol, "Ullmann's Encyclopedia of Industrial Chemistry," Weinheim : Wiley-VCH Verlag/Gmbh & Co; 2002.
[21] Meyers RA, "Handbook of synfuels Technology," New York : McGraw-Hill; 1984.
[22] Chu YZ, Qin LP, "Optimum design of methanol distillation process and columns," Chem Ind Eng Prog 2008;(27):1659-62.
[23] Juntao Zhang, Shengrong Liang, Xiao Feng, "A novel multi-effect methanol distillation process," Chem Eng Pro 2010;(49):1031-37.
[24] Jinsheng Sun, Fan Wang, Tingting Ma, Hong Gao, Peng Wu, Lili Liu, "Energy and exergy analysis of a five-column methanol distillation scheme," Energy2012;45:696-703.
[25] J. S. Sun, P. Wu, Y. H. Wang, H. Gao, L.L. Liu.CN Patent102039058 A. 2011.
[26] J. S. Sun, P. Wu, Y. H. Wang, Y. X. Su, H. Gao, J. Li, D. Yan. CN Patent 101798322 A. 2010.
[27] Michele Mascia, Francesca Ferrara, Annalisa Vacca, Giuseppe Tola, Massimiliano Errico, "Design of heat integrated distillation systems for a light ends separation plant," Applied Thermal Engineering 2007;27:1205-1211.
[28] Olga A. Flores, J. Carlos Ca´ rdenas, Salvador Herna´ ndez, Vicente Rico-Ramı´rez, "Thermodynamic Analysis of Thermally Coupled Distillation Sequences," Ind. Eng. Chem. Res. 2003;42:5940-5945.
[29] Szargut J, "Exergy Analysis of Thermal, Chemical and Metallurgical Processes," New York: Hemisphere Publishing Corporation;1988:55-152.

(45n) In Situ Transesterification of Jatropha Curcas Seed Particles with Alkaline Phase Transfer Catalyst

Subbarao Duvvuri[1], Shuhaimi Mahadzir[2] and Sintayehu Hailegiorgis[2], (1)Department of Chemical Engineering, Universiti Teknologi PETRONAS, Tronoh, Malaysia, (2)Chemical Engineering, Universiti Teknologi PETRONAS, Tronoh, Malaysia

Economic viability of biodiesel technology is at stake due to the high cost of edible oil feedstock and slow reaction rates due to limited solubility of oil in methanol. In situ transesterification of non edible oil seed particles can minimize the cost of feedstock. Microwave pretreatment of seeds can make oil molecules more reactive. Use of phase transfer catalysts can enhance the reaction rates. In the present work, these three concepts are utilized together to investigate alkaline in situ transesterification of microwave pretreated jatropha curcas seed particles in presence of benzyltrimethylammonium hydroxide (BTMAOH) as a phase transfer catalyst (PTC). It was observed that addition of PTC to alkaline catalyst has improved conversion and reduced reaction time. Microwave pretreatment of jatropha seed particles reduced reaction time drastically. The quality of biodiesel produced was within the limits of the international standard requirements of ATME D6751 and EN-14214. Thus, combined use of microwave pretreatment of seed particles, phase transfer catalysis in the presence of alkaline catalyst can increase the rate of in situ transesterification and product yield.

(45o) Catalyzed-Assisted Manufacture of Olefins (CAMOL): Updated for Use in Naphtha Service

Maximilian Walter, Petrochemicals, BASF SE, Ludwigshafen, Germany and **Suzanne Rech**, BASF Qtech, BASF Corporation, Iselin, NJ

Since the commercialization of the olefins plant, operators have searched for ways to be more competitive. One area always targeted includes keeping their furnaces in operation longer between decoking steps. Over the years, many anti-coking methods/technologies have become commercially available using different approaches to prevent coils from coking. One of the latest available technologies is the novel CAMOL™ (Catalyzed-assisted Manufacture of Olefins) Technology from BASF Qtech. As reported in the introduction of the technology to the EPC in 2008 & 2010, the CAMOL™technology differs from other available anti-coking technologies in that the technology not only eliminates the catalytic (or filamentous) coke make through a physical barrier, but also gasifies the thermal (or amorphous) coke for an even greater coking resistance. The coating process has generated a coating with almost identical thermo-mechanical properties as the base metal substrate (alloy). This allows bonding to the base metal similar to being welded with a minimum risk of spallation when going through thermal cycling. Commercially, the CAMOL Technology has been available since 2006 and is now in operation in furnaces of various designs.

With CAMOL™, operators have experienced extended run lengths, reduced energy requirement and GHG emissions, lower average TMT operating temperatures, reduced DMDS consumption, and less carburization. In addition, the technology has exhibited high operating temperature stability (>1130°C/ 2065°F).

At BASF SE in Ludwigshafen, Germany, the technology has been in operation since early 2010 in mostly naphtha service. After overcoming some initial challenges associated with a catalytic surface technology, such as reactivating the catalytic species after decoking, the coil is now successfully operating with run lengths significantly higher than uncoated coils. This paper will report on the results and learnings from the experiences gathered over the past three years along with the advancements made since the last report in 2010.

(45p) Extending Furnace Run Length and Coil Life by Optimizing Decoking

Vinod Mishra, Ingenero, Mumbai, India and **James Brigman**, Ingenero, Houston, TX

Abstract:

In the Ethylene manufacturing process, coke deposited in the radiant coils and Transfer Line Exchangers (TLEs) during normal operation is removed periodically by decoking the furnaces using a mixture of steam and air. There is a significant variation in operating parameters adopted by different plants for decoking. These are often either too conservative or aggressive, leading to production loss from increased off-stream time, TLE limitation and deterioration in coil life. This paper describes how the decoking procedure was modified based on furnace and TLE operating conditions and the decoking steps changed to obtain the best possible Ethylene yield, without adversely impacting coil life and TLE performance. This optimization has now been done in several major ethylene facilities and furnace run length has more than doubled and production has significantly increased.

SOLVING OPERATIONAL PROBLEMS IN A NAPHTHA SPLITTER

Flávio Weissheimer, GE Water & Process Technologies do Brasil Ltda, São Paulo/Brasil
Giorgio Franceschetti, Braskem, Santo André/Brasil
Rafael Henrique Camparin, GE Water & Process Technologies do Brasil Ltda, São Paulo/Brasil
Patrick Lucas, GE Water & Process Technologies, The Woodlands/US

Abstract

This paper sets out how the polymerization process of Reboilers of the (A-100) Crude Naphtha Fractionation Unit of Braskem UNIB 3 ABC can be controlled by applying polymerization inhibitors developed by GE Water & Process Technologies Ltda and Braskem´s engineering solutions.

These heat exchangers have a high degree of polymerization because of the content of olefinic compounds present in the feed composition which, when in contact with high temperatures in the heat exchange areas of reboilers, tend to polymerize. The joint solution of GE technology and Braskem engineering solutions applied to this system ensures regular operation of such equipment for periods of time greater than those hitherto recorded without manifesting loss of heat transfer efficiency.

In the 2012 shutdown while the opening of these reboilers was being inspected, it was observed that pipes had not been clogged nor was surface fouling present, which is a very positive state when compared to the historical records of previous campaigns.

The strategy adopted by GE/Braskem to ensure a longer campaign for these pieces of equipment was based on the quality control of feeding by adjusting dosages and monitoring performance using statistical analysis tools.

The use of statistical models based on Multivariable Regression Analysis (MRA) to evaluate performance, which is validated by inspecting reboilers internally, has been an efficient way to determine the degree to which the system has been fouled and also to schedule shut-downs for cleaning the mechanical parts.

Description of the Process

The (A-100) Naphta Fractionation Unit includes two distillation columns, the objective of which is to separate the light (C_6), medium (C_6 - C_8) and heavy (C_{9+}) cuts, included in the feed supply of crude naphtha.

The light cut and a portion of the heavy cut are sent to pyrolysis furnaces so as to produce ethylene. The medium cut is sent to the Hydrotreating and Catalytic Reforming Unit, with the objective of increasing the aromaticity of the system so as to produce BTX.

The A-100 processes 100 tons/h of crude naphtha, and approximately 35% of this current represents the average cut. Figure 1 below shows the schematic flow diagram of this fractionation process.

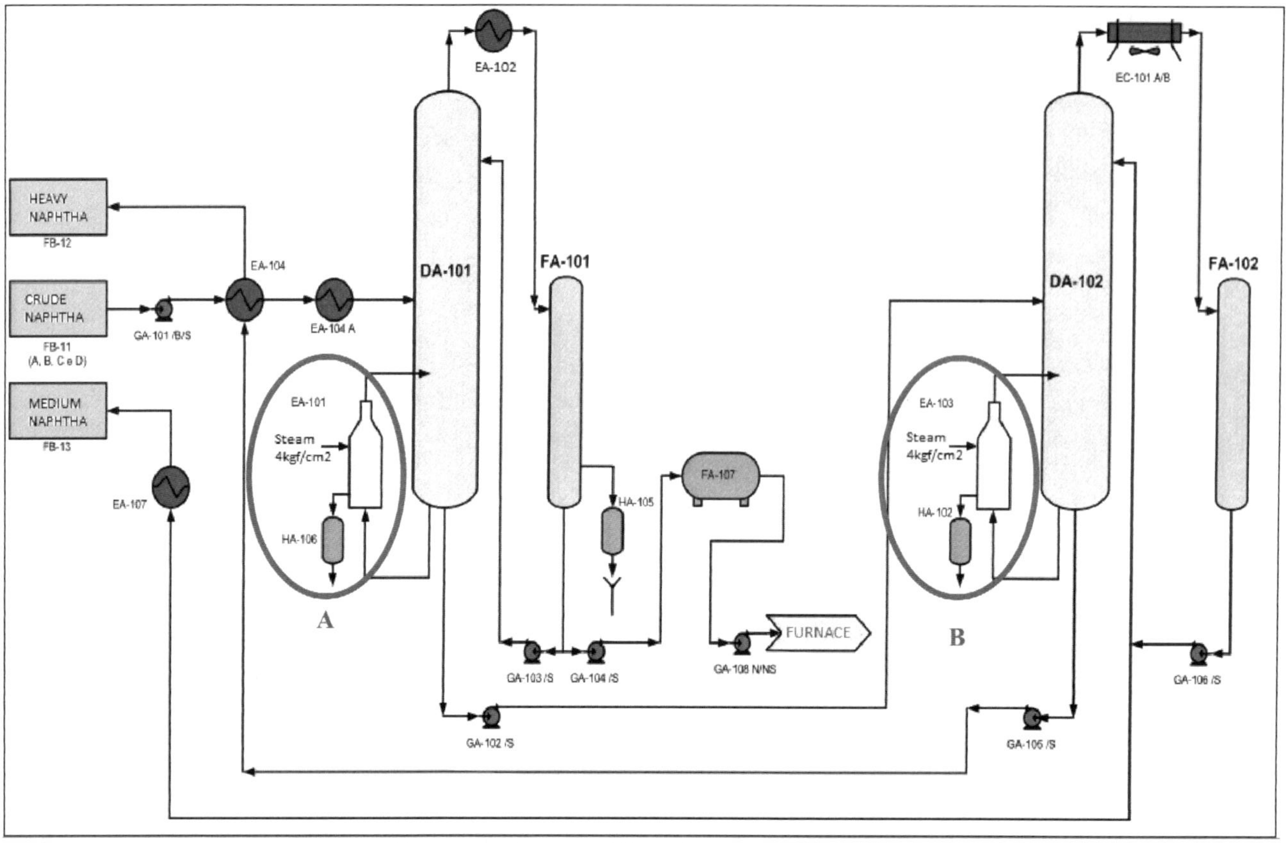

Figure 1. Process Flow Diagram (A-100)

The Reboilers, indicated by the letters A and B in Figure 1 above, operate with an injection of low steam (4kgf/cm2) to promote vaporization of the bottom stream of the A-100 distillation columns. These devices operate over a temperature range of about 150°C, which is sufficient to promote the polymerization of certain olefinic compounds present in the feed of crude naphtha.

Poor performance of these Reboilers can lead to production stoppages for maintenance and cleaning, which very often affects up to five days of production. The possibility of raising the times of campaigns and predicting the degree of fouling in downstream units helps to minimize production losses due to unscheduled stoppages.

This study will discuss the results of Reboiler A, which is the most critical equipment, with a view to synthesizing data, but gains in equipment B were also significant.

Evaluation System

Historical Record of the Campaign

Reboiler A was installed in 1972 but examination of the historical records did not reveal that the sheaf of tubes had ever been replaced. This exchanger has currently 11 tubes plugged into it.

The Campaign times of this equipment are shown in Table 1, with the last campaign being highlighted in blue so as to set it apart from the other ones. Table 1 gives a summary in the order of the timings of the campaigns, the average load processed in the period, the total load processed, the reason for operational stoppages etc.

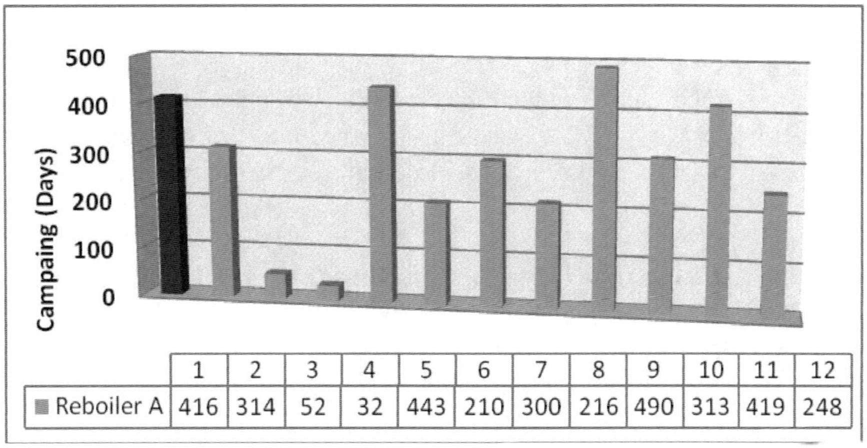

Figure 2. Campaign of the Reboiler A

Table 1. Historical Record of Reboiler A Campaigns

Date	Campaign *(days)*	Capacity *(ton/h)*	Capacity *(ton/campaign)*	Occurrence	Tubes Clogged *(%)*
13/06/12	416.0			Programmed Shutdown	0
17/03/11	314.0	93.1	701601.6	Heat exchange performance	30
03/05/10	52.0	90.8	113318.4	Stuck tube	-
11/03/10	32.0	71.1	54604.8	Heat exchange performance	50
09/02/10	443.0	72.0	765716.6	Heat exchange performance	20
16/11/08	210.0	98.1	494575.2	Heat exchange performance	-
16/04/08	300.0	104.0	748800.0	Programmed Shutdown	-
16/06/07	216.0	92.8	481075.2	Programmed Shutdown	-
10/11/06	490.0	-	-	Programmed Shutdown	-
30/06/05	313.0	-	-	Stuck tube	-
17/08/04	419.0	-	-	-	-
18/06/03	248.0	-	-	-	-
10/10/02	-	-	-	-	-

Conditions of Opening

In this review, a comparison is presented between two operational shut-downs of A-100, which occurred in March 2011 and May 2012, in order to observe the improvement in performance from the last campaign associated with the change in the GE Treatment strategy.

Campaign I: Production shutdown on 03/2011

Table 2 below summarizes the features observed during the inspection of the state of the inlet of this piece of this equipment during the shutdown of March 2011:

Table 2. Internal inspection of Reboiler A in 03/2011

Bottom Cover	Surface residue
Bottom Mirror	No deposits or internal blockage of the tubes were observed.
Top Mirror	Approximately 30% of the tubes were clogged, especially in the region of the vapor inlet, which has higher temperatures. Magnetic and non-magnetic deposits.

May/2011

Figure 3. The mirror of Reboiler A before it was cleaned.

May/2011

Figure 4. The mirror of Reboiler A at the vapor inlet before it was cleaned.

Note: It is difficult to carry out internal cleansing by Hydro jetting (pump pressure during cleaning is 20,000 psi). This reboiler has 11 tubes plugged into it, but none of them are plugged in during this shutdown.

Campaign II: Production shutdown on 03/2011

Table 3 below summarizes the features observed during the inspection of the state of the inlet of this piece of equipment during the shutdown in May 2012:

Table 3. Internal inspection of Reboiler A in 05/2012

Bottom Cover	Surface residue
Bottom Mirror	No deposits or internal blockage of the tubes were observed.
Top Mirror	Clogged tubes were not identified, not even in the region of the inlet vapor. There was only a thin layer of deposit that was easy to remove and which was evenly distributed over the upper surface of the mirror.
Spool	A thin deposit on the inner surface of the caisson.

May/2012

Figure 5. The mirror of Reboiler A before it was cleaned.

May/2012

Figure 6. The mirror of Reboiler A before it was cleaned.

Note: Hydro jetting was done quickly, without any obstructions due to deposits having been identified. This equipment has 11 tubes plugged into it, but none of them were during this shutdown.

Analysis of Deposits

The characterization of the deposits found in Reboiler A showed an 86.0% loss on ignition, thus indicating the strong presence of organic compounds in their composition, with the functional groups of carboxylic acid and ester standing out. The inorganic analysis of the sample indicated a high concentration of iron oxides and also sulfur oxides in the form of sulfides on the cover and spool of this reboiler. This is summarized in Table 4 below:

Table 4. Analysis of the Deposits of Reboiler A

Organic Compounds - LOI (%)	86.0
Water of hydration	Smaller
Sulfur compounds	Smaller
Carboxylic Acid and Ester	Greater
Dichloromethane soluble (%)	2.0
Elemental Sulfur	Greater
Dichloromethane insoluble (%)	98.0
Water of hydration	Smaller
Inorganics	Smaller
Carboxylic Acid and Ester	Greater
Composition of Dichloromethane insolubles (%)	
Carbon	56.0
Hydrogen	5.0
Nitrogen	3.0
Sulfur	8.0
Inorganic Compounds (%)	
Iron, $Fe_2O_3 + Fe_3O_4$	82.0
Sulfide, S^{2-}	16.0
Phosphate, P_2O_5	1.0
Sodium, Na_2O	1.0

Evaluation of Performance

The evaluation of performance of Reboiler A was prepared using the MRA tool, which is a GE good practice to predict fouling in critical systems. This statistical tool enables a dependent variable to be determined, in this case the percentage of opening the steam valve, by means of the numerical regression of the other operational variables of the column. This method generates a mathematical equation to predict the percentage of inlet steam when the equipment is clean, given that the equation obtained by regression includes data of this condition.

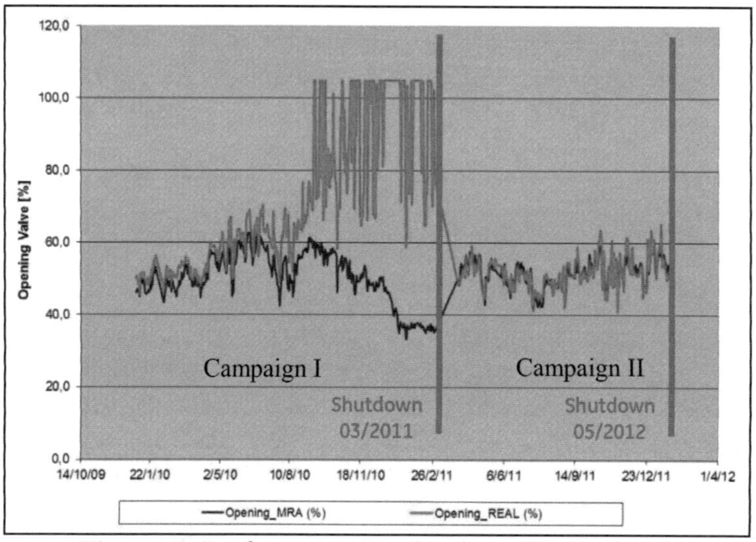

Figure 7. Performance Evaluation of Reboiler A

In Figure 7 the dark blue line represents the percentage of inlet steam when the equipment is clean, as per the equation defined by the MRA. The pink line represents the actual percentage of inlet steam. The sharp upward shift evidenced in the period between November 2010 and March 2011 shows the fouling of the system during Campaign I (prior to the 2011 operational shutdown).

After the 2011 shutdown, it can be seen that the two curves follow the same trend, without a shift, indicating that the unit has the equivalent performance of a clean piece of equipment. The change in the GE treatment strategy, taking the control of the dosages of the product Petroflo 20Y25 into account, was seen to have been effective in increasing the campaign time of this equipment.

Conclusion

The change of strategy in applying dispersant and controlling the dosages after the March 2011 shutdown, presented positive results, given the states of the inlet seen during the inspection of reboilers A and B in May 2012..

The low degree of fouling of these reboilers evidenced in the 2012 shutdown led to the production process being considered reliable, as the A-100 reboilers are no longer the pieces of equipment responsible for the loading limit of the unit. The downtime during the cleaning of the mechanical parts of the reboilers was about 10 times less than that observed in the previous shutdown, thus reducing Braskem´s maintenance costs.

In addition, the MRA statistical tool was able to produce a reliable representation of the degree of fouling of the system. This tool can be effective for scheduling shutdowns more than 03 months in advance.

References

1. Setal Lummus, "Manual de Operação da Unidade de Processamento de Nafta Bruta (Área 100)", Santo André, Brazil.
2. Ponce de Leon, Carolina; Weissheimer, Flavio and Camparin, Rafael H. (2011), "Proposta Técnica de Fornecimento de Especialidades Químicas, Equipamentos e Serviços para Tratamento Químico para a Unidade de Processamento de Nafta Bruta (Área 100)", Santo André, Brazil.
3. Kleinbaum, David G., Kupper, Lawrence L., Muller, Keith E. (2008) "Applied Regression Analysis and Other Multivariable Methods", Belmont, USA.

13AIChE
Spring Meeting, San Antonio, TX

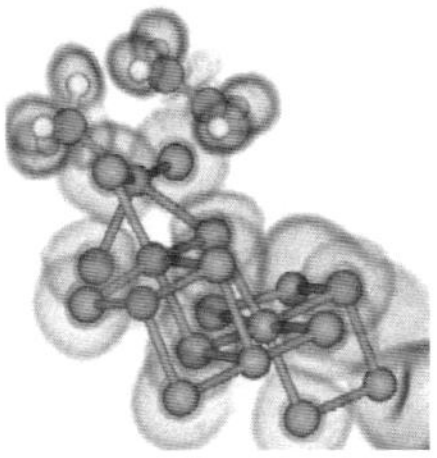

SmartProcess® Application Strategy Achieves Faster and More Efficient Control of an Ethylene Complex

Patrick Truesdale
Principal Consultant
Industry Solution Group
Emerson Process Management

April 29, 2013
San Antonio, TX

EMERSON
Process Management

Speaker Introduction

- Principal Consultant
  Emerson Process Systems & Solutions
- BS Electrical Eng, PE
- 40+ Years experience with Refinery & Petrochem Process Control
- Joined Emerson in 2007
- Previous employers: duPont, Bonner & Moore, SETPOINT, AspenTech, Comcept Technology

Patrick Truesdale

E-mail: patrick.truesdale@emerson.com

13AIChE
Spring Meeting, San Antonio, TX

EMERSON
Process Management

Agenda

- Performance Package Review
- SmartProcess Ethylene Overview
  - Furnaces
  - Distillation
- Benefits

SmartProcess Ethylene Packages

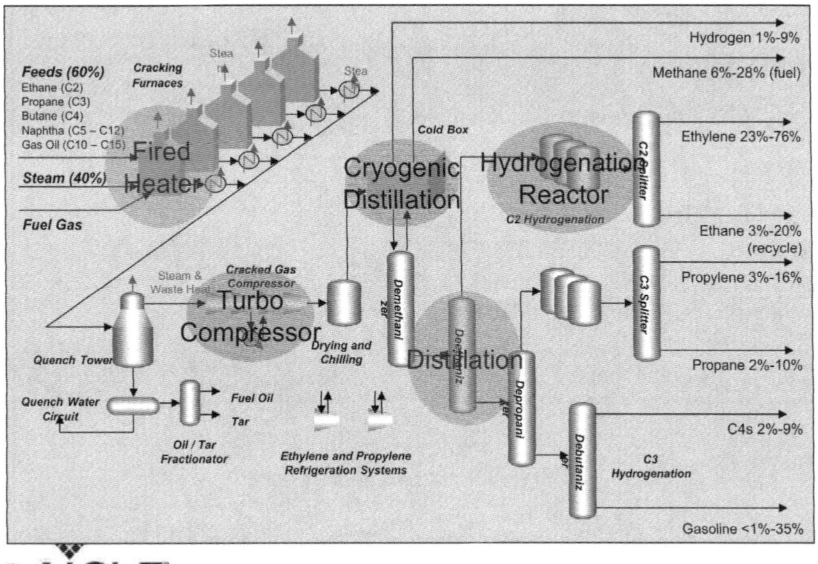

SmartProcess Scope

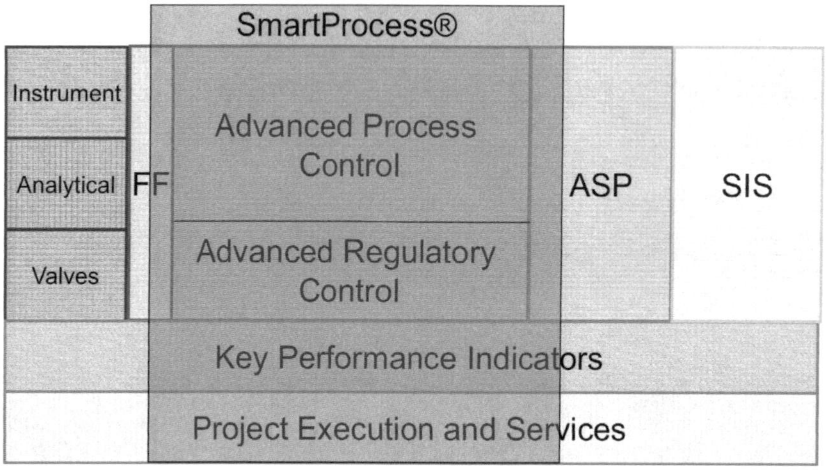

Basic Ethylene Furnace Variables

Key Operating Objectives:

- Constant severity
- Maintain HC/Stm ratios
- Min excess air
- Maintain within constraints
- Max charge (if needed)

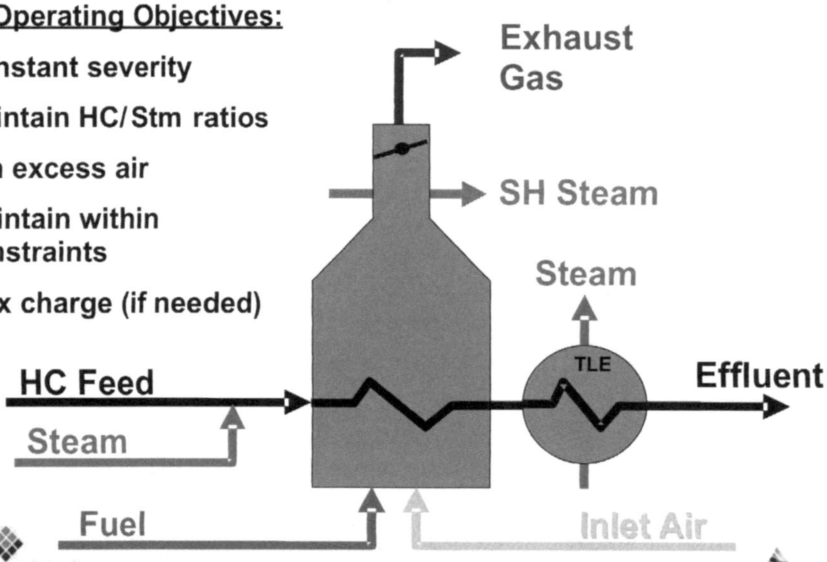

SmartProcess Furnace Module

- Design allows for different furnace configurations
  - Process Zones
  - Firing Zones
  - Multiple TLE analyzers
- Advanced Regulatory
  - Charge Control
  - Steam Control
  - Severity Calculation
  - Combustion Control
  - KPI's
- Advanced
  - Pass Balancing
  - Heat Balancing
  - Severity Control
  - O2 Control

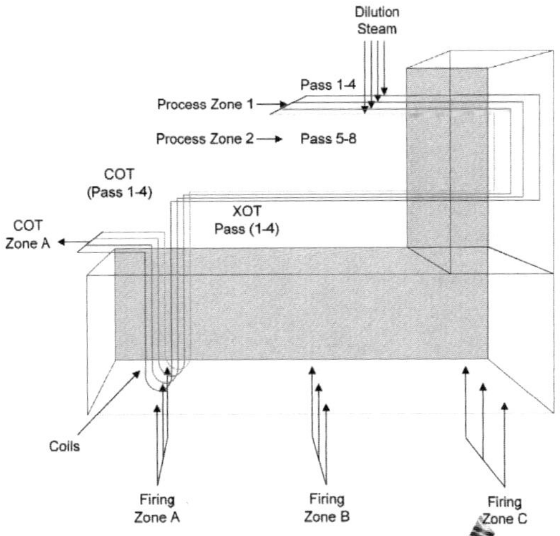

13AIChE Spring Meeting, San Antonio, TX

EMERSON. Process Management

Standard Furnace Module Design

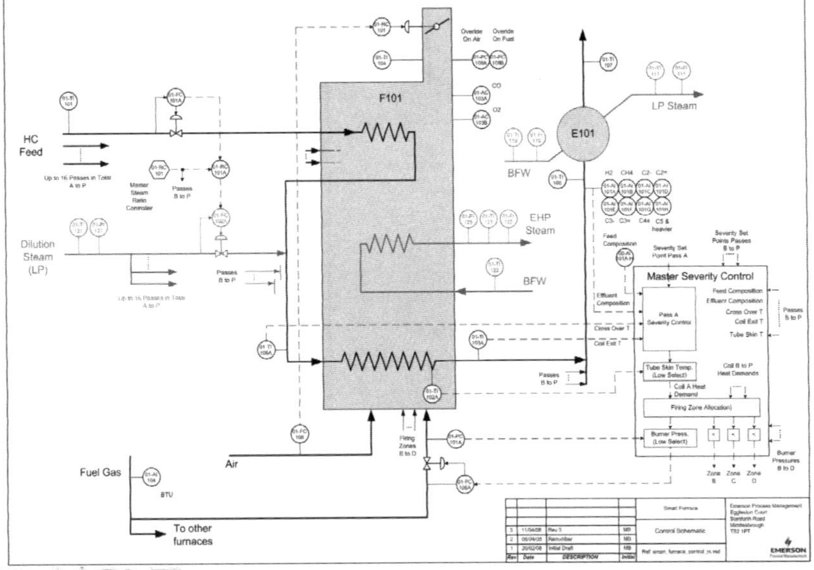

13AIChE Spring Meeting, San Antonio, TX

EMERSON. Process Management

Charge Rate Control

Functions:
- Pass Balancing
- Constraints
- Feed Maximization

Functions:
- Ramping
- Limits & Alarms
- Multiple Feed Types
- Furnace Status
- Pass Offsets

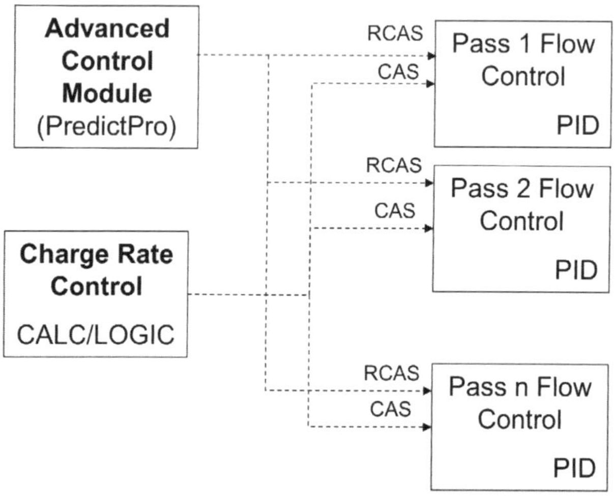

13 AIChE
Spring Meeting, San Antonio, TX

EMERSON. Process Management

Steam Ratio Control

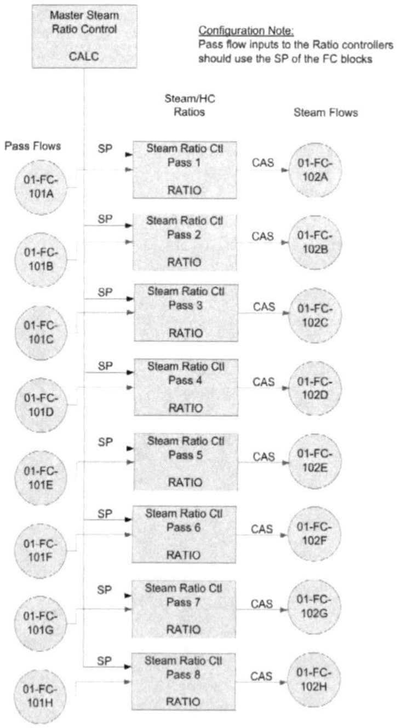

13 AIChE
Spring Meeting, San Antonio, TX

EMERSON. Process Management

Severity Control

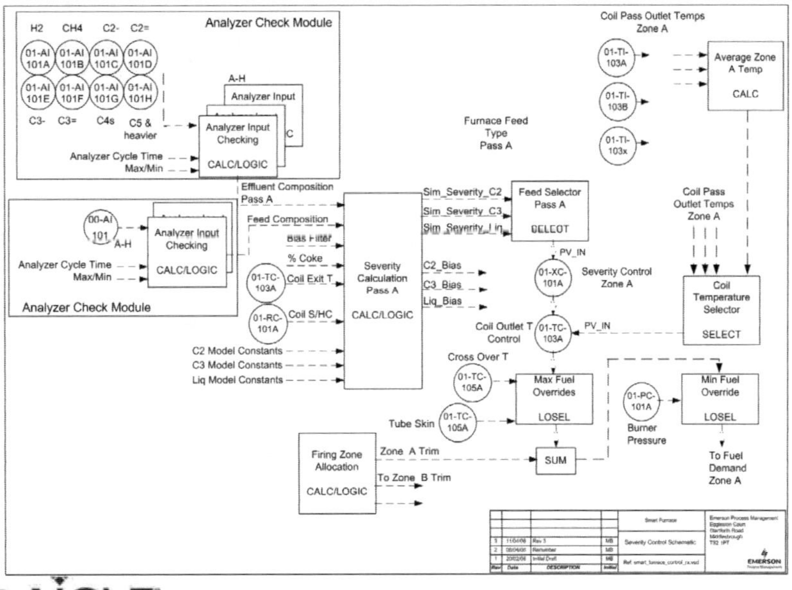

Automated Furnace Sequences

- Standard module templates to be used for Furnace Startup, Shutdown and Decoking sequences.
- Utilizes preconfigured Sequence Function Charts (SFC) running in DeltaV controllers or App Station
- Standard sequence functions can be customized to match site requirements
- Manual operator steps with confirmation can be mixed with automated steps
- Sequence execution control graphics and faceplate templates

Key Performance Calculations

- Furnace efficiency
  - Heat loss method
- Furnace Duty
- TLE Duty, Fouling factor
- Stack gas estimated composition
- Stack gas daily total tons exhaust for CO, CO2, S, N, etc.
- Stack gas NOx emissions (optional – based on CEMS)

13 AIChE Spring Meeting, San Antonio, TX

EMERSON. Process Management

Distillation - Classic Multivariable Problem

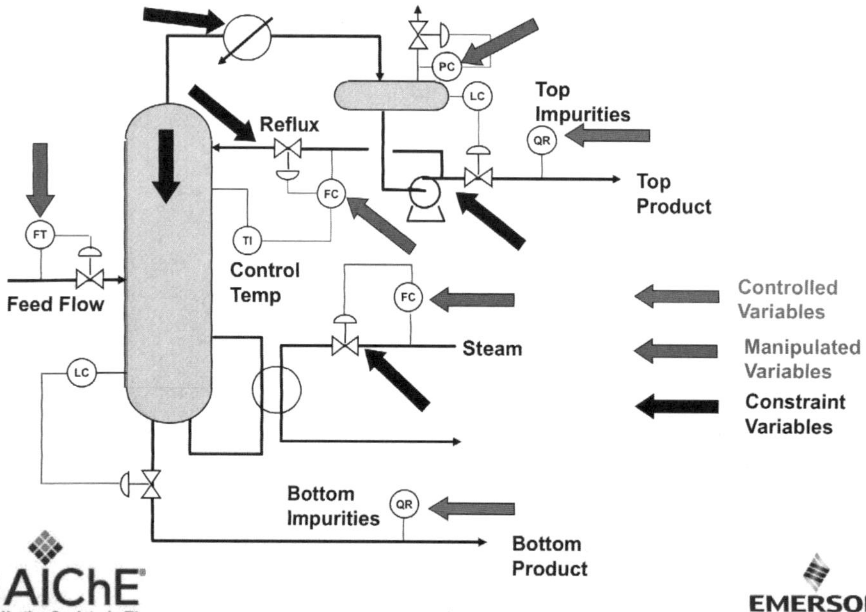

13 AIChE Spring Meeting, San Antonio, TX

EMERSON. Process Management

Distillation Control Module

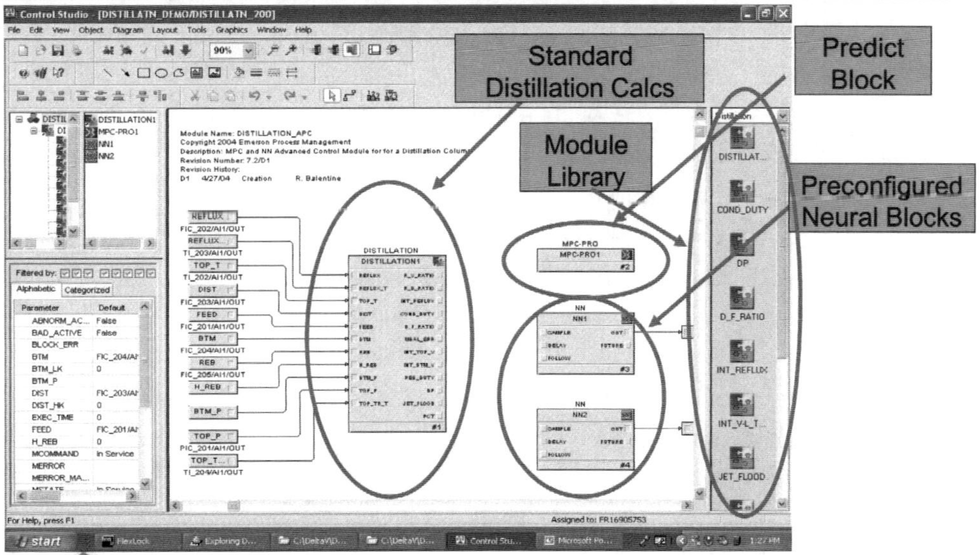

SmartProcess Distillation

- DeltaV PredictPro with up to 4 MV's
- DeltaV Neural for up to 2 product qualities
- Standard set of calculations & constraints
  - Internal reflux
  - Pressure compensated temperatures
  - Jet flooding
  - Reboiler & condenser duty
- Engineering Services to design, implement and commission

SmartProcess Applications

- Licensed industry applications for common processes that are:
  - Pre-engineered, tested, documented
  - Designed by APC Consultants with extensive practical experience
  - Supported & enhanced
- Bundled combination of:
  - DeltaV libraries: composite blocks, modules & templates
  - DeltaV simulation and demonstration example
  - APC tools (e.g. Predict, Neural)
  - Application code (e.g. VB, .Net)
  - Other databases and apps (e.g. SQL*Server, Excel, TankMaster)
  - Sample DeltaV Graphics
  - Documentation
- Templates to be used to configure for specific site requirements
- May be implemented by Emerson or Customer

Benefits

Ethylene Furnace High Temperature Imaging Cameras to Monitor Burner Flames ,Tube Temperature and Tube Failure Detection

Thomas M. Canty, P.E.
President
J.M. Canty Inc
Buffalo, NY USA

Abstract

The development of explosion proof, high temperature imaging cameras which operate in ethylene furnaces without cooling air now allows monitoring of ethylene furnaces to improve energy/operating efficiency along with providing important safety benefits. Imaging cameras work off the visual wavelength which provides 2 dimensional view and temperature measurement of the tube temperature which is more accurate than thermocouples and IR probes. In addition visual images of the furnace burners lets you detect flame detachment, smoking burners and flame shape in order to increase burner efficiency and save energy. The safety aspects of the Camera imaging systems is significant as it allows visual detection of pin hole leaks and tube failure in the furnaces.

Introduction

Ethylene furnaces, or any type of cracking unit, rely on thermal management in order to make product efficiently and to operate in a safe manner. Having 'eyes' in the furnace can help to identify problems in their early stages of development. High temperature camera technology is able to survive the environment and give a wide angle view of the tubes from which temperature and tube blockage may be determined. Achievable benefits from a camera system include:

1. Temperature analysis capacity for all areas of the image, not just single point. Multiple zone capability for user customization of analysis.
2. Flame detection and size measurement
3. Detection of tube movement and ability for software to automatically move measurement zones as tubes distorts.

Imaging Fundamentals

Basics of an imaging system include a high resolution sensor, proper illumination, appropriate hardware for the process and a robust software analysis package. For this particular application illumination is created by the process and need only be managed optimally for best results. A high density pixel array yields resolution of the image and measurement as well. The hardware and software are the more challenging aspects of the application and will be discussed in some detail.

Hardware

The design of the camera system is unique in that it's internally exposed lens assemblies are **designed to survive without cooling air**. When installed properly, the combination of system materials and heat transfer coefficients at the camera tip limit heat transfer to the system and allow it to run hot without chance of damage. This feature, when system is installed in a clean, gas burning environment, does not disrupt the temperature or air balances of the furnace because no cooling air is being introduced to the furnace. In addition, the design includes a disposable, quarts shield which protects the process exposed lenses from direct contact with the internal atmosphere. Should any particulate contact the system, the quartz shield is there to protect the optical lenses. Replacement of the protective quartz shield is inexpensive. Replacement of damaged lenses is very expensive and so the Canty design guards against that possibility.

Internally, the system is a two piece assembly consisting of the internal insertion lens and the external camera package. As mentioned earlier in this document, Canty employs a fused glass window as part of the electronic housing package. This has a dual purpose; Firstly, the camera housing assembly is a separate entity which carries WP and EXP ratings. Secondly, the camera housing bolts to the internal lens system and transmits the optical rays through the fused glass window to the CCD. This mounting also creates a hermetic barrier between the process and the electronic housing thereby preventing migration of hot, process gasses outside the pressure boundary of the system.

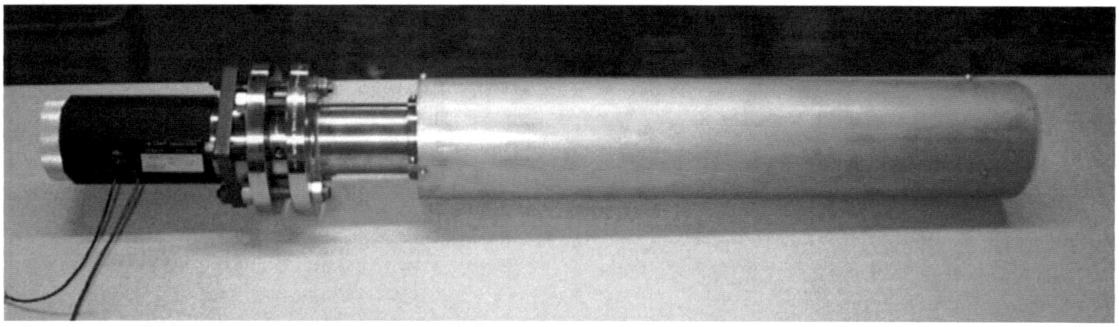

Figure 1. High Temp Camera Assembly

The camera electronics come with a full array of controls that allow it to be optimized to the software function to achieve accurate, consistent data output.

As previously mentioned, the installation is an important feature of the design. When installed correctly, the camera sits inside the refractory (see Figure 2) where its exposure to the process environment is minimized thus minimizing heat transfer into the internal lens.

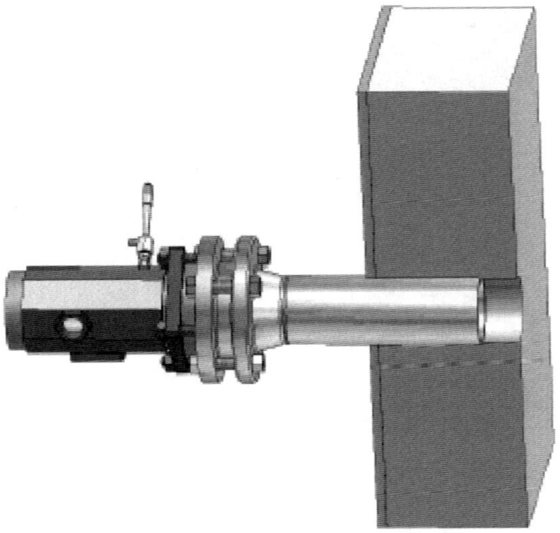

Recessed into Refractory

Figure 2. Camera Installation Schematic

Software

With respect to software design, the Canty software is fully flexible in analyzing any portion of the captured image. Multiple analysis zones can be drawn to monitor discreet areas of the view and each zone can run its own particular algorithm depending on need. The benefit here is that multiple tubes can be continually surveyed for changes and trends. The analysis data is also not single point, but is comprised of all the data points in the analysis zone. By choosing proper software filters various information can be obtained from the full array of data such as hot or cold spots. This data can also be archived in total or as snap shots at determined increments of time. The on screen data output can also be updated in the short or near term and can thus be used to judge what is happening in the furnace over time with just a glance at the screen. For instance, using the histogram feature with a 1x daily output a long term trend can be presented on screen in real time.

The basic theory of temperature measurement is based on the radiative emissions of a body at temperature – as the body temperature increases, the radiation curve shifts towards the shorter wavelengths and that is why as a body heats up the eye cannot detect it until it reaches 1000F which is the approximate visible point where emitted radiation is dark red in color. As the temperature increases, the color graduates to bright red, then orange then toward white. A pyrometer functions by measuring a narrow band of wavelength intensity and comparing it to the intensity of a black body to determine the temperature. From the graph below it can be seen that if a single wavelength is monitored the intensity of that measurement changes as the body temperature changes (curve shifts) which then corresponds to a different temperature on the calibration scale.

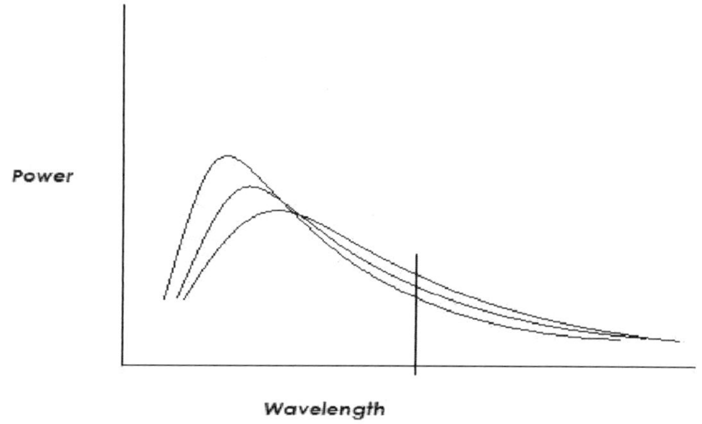

Figure 3. Power Curve at Varying Temperatures

The same is true of a camera system measuring the increasing intensity of an object as it radiates from approx 1000 F (dim, dark red color) on up the temperature scale to 3000+ F for instance which would be a bright white. From the graph above one can see that as color goes from red toward white the peak intensity is always increasing and it is the measure of this intensity which indicates the object temperature. Using an RGB camera the measurement can also be made by monitoring the individual color components of the signal impinging on the chip. For industrial uses, the measurement of intensity usually suffices.

Standard instrumentation used in industry include thermocouple type devices (contact) and thermal imaging and pyrometer systems (non contact). In severe temperatures and atmospheres, the contact devices deteriorate quickly and provide gradually less reliable information leaving the operator running blind. They also create a discontinuity on the surface they are welded to which, in the case of a furnace tube, can lead to premature tube failure.

Non contact systems have difficulties with calibration drift and emissivity adjustments to account for the fact that many items being measured are not black body emitters. In addition, these technologies are single point measurement systems and are best used as hand held devices which can be difficult to use and provide subjective measurement results based on operator preferences for targeting.

Vision technology offers several benefits these other technologies do not. The visual aspect of the system allows the operator to view the process and confirm the readings received. It also allows flexibility to set up measurement zones on screen at any time to assess any or all areas of the process in view should a change be required. Since vision systems are generally fixed installations, an added benefit is that the information is taken from the same process points and features within a given measurement zone at all times which means there is no error in sighting of the instrument.

The vision system is also unique in that it does not suffer calibration drift as almost all instruments do. Once the calibration is set the readings repeat themselves within a couple degrees F. The calibration is based on thermocouple or pyrometer readings initially, however after that calibration is finished it does not have to be redone in the future as long as the system remains in its location.

Process history and performance can also be best documented by a vision system. First off, the camera can analyze any area of process that is within the systems view. This affords the operator the capability of viewing many points of interest instead of just single point temperature readings. As an example, if the system is monitoring a tube bank it can assess the temperature of the tubes for the full length of view. If there is a local flow problem in a tube it will become overheated in one section. The camera can locate that section and indicate its temperature rise over the rest of the tube or adjacent tubes. This information allows the tube location to be identified as well as the position in the tube that has encountered the upset condition. Over time this historical data, recorded in video or still image format by the control software, may be significant in improving the process. In addition to the ability to monitor the full area of the view, the full area of view can also be segmented into multiple zones of measurement, i.e. each tube having its own measurement zone for automated process control capabilities. This capability allows the tube to be operated at optimum efficiency since the addition of excess heat can be avoided thereby reducing energy draw to operate the furnace.

Thermal Analysis

Tube Temperature

The following figures illustrate the temperature analysis capability for tubes in a live test situation. Measurement zones are placed over the tubes (figure 4) as the furnace is heated and cooled. The temperature plots are shown in figures 5 and 6. The results are as expected and uniform from tube to tube, although a small number are included in this demonstration.

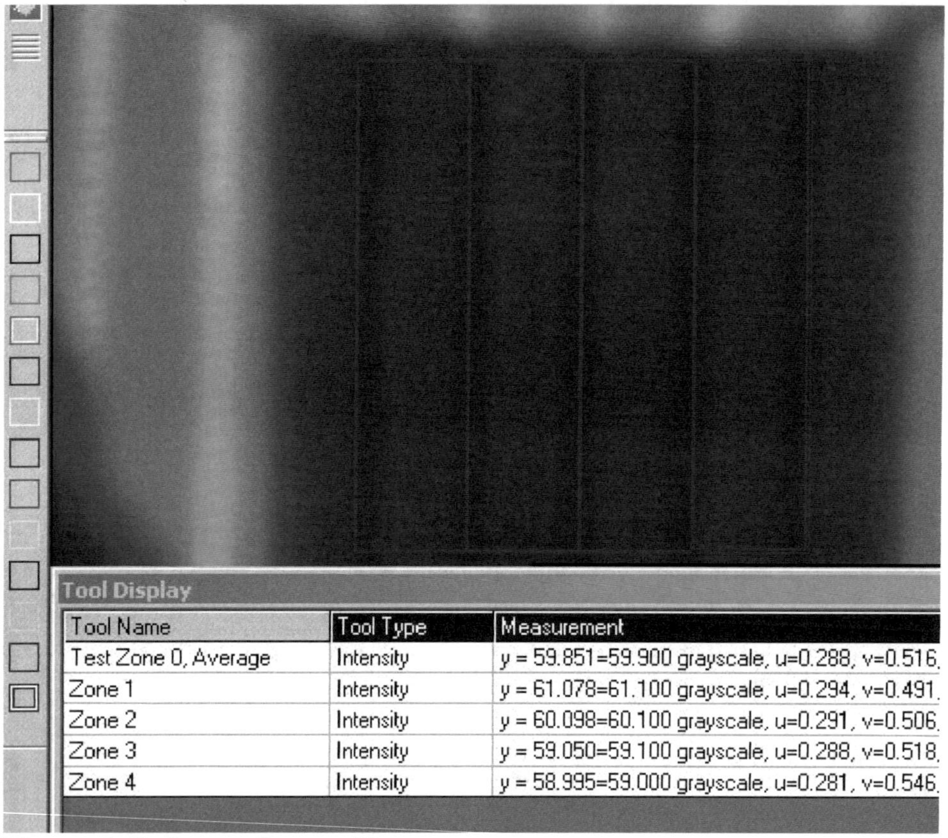

Figure 4. Software Screen Shot

As thermal loads are fluctuated up and down the following graphs track those changes. The cooling cycles are left to natural convection and thus are not linear in nature.

For the following curves, the vertical scale is gray scale intensity which is measured 0 – 255 (pitch black to bright white). The hotter the tube, the higher, or more intense, the gray scale reading. This value in process would be calibrated to the temperature. The horizontal scale is the image number, or time.

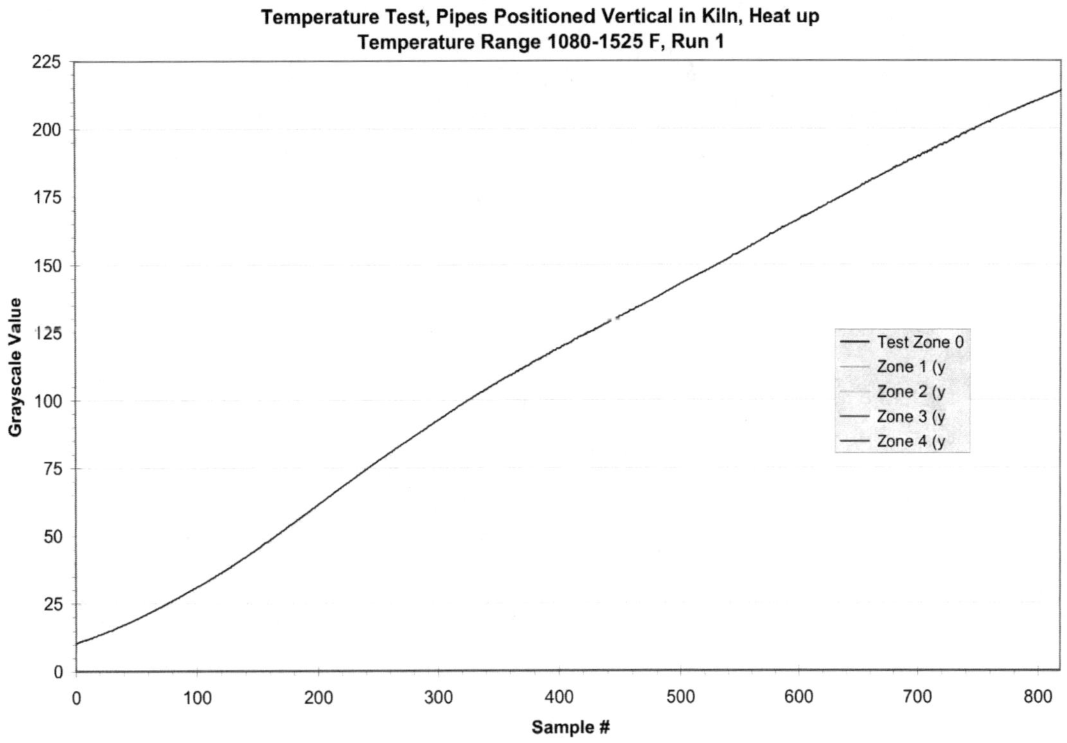

Figure 5. Heating Cycle

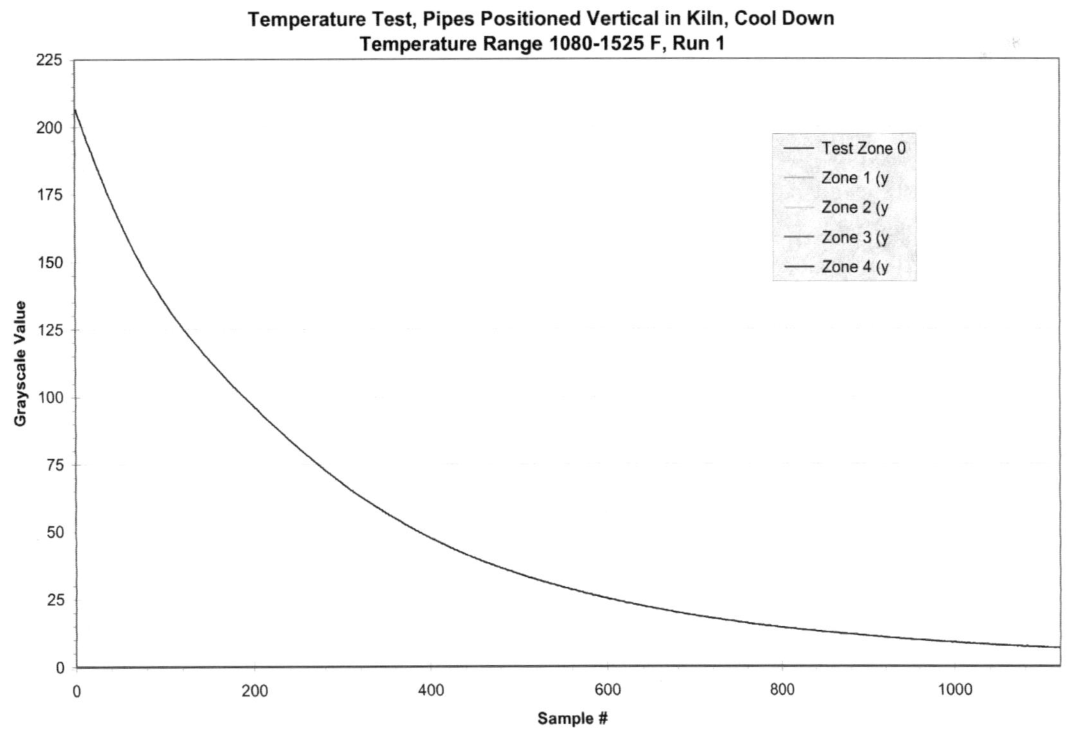

Figure 6. Cooling Cycle

Flame Measurement
Flame temperature can be monitored in the same way as described previously, however the flame size, or presence, can also be monitored. Figure 7 shows a live flame and Figure 8 shows a digital capture. Computing the flame area, or shape, can be done to track its changes over time.

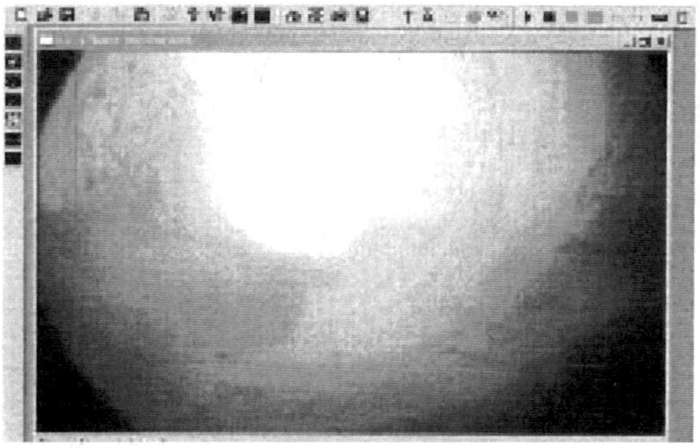

Figure 7. Kiln Flame (Live)

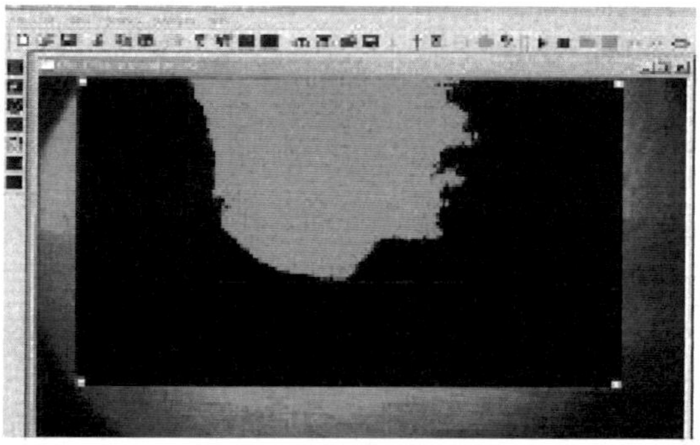

Figure 8. Kiln Flame (Digital)

Conclusion

Vision Technology presents an opportunity to increase effectiveness of furnace monitoring through hardware and software capabilities. The thermal design at the core of the hardware enables the system to provide the operator a view into the furnace and feed back on its thermal performance over time. This data can assist in detecting building upset conditions, but also provide a record for study to assist in improving the operational method of the furnace.

(45v) Straighten up! Non-Linear Compensation Using Feed Forward Control

Barbara Hamilton, Process Systems and Solutions, Emerson Process Management, Hampden , ME

Non Linear processes occur throughout the petrochemical plant. While the PID loop is surprisingly robust for most situations, there are cases where compensation of these non-linearity's is critical for stable control. Similarly, disturbances can cause major disruptions in control stability.

A method of feed forward control, using data based models, can provide robust control throughout the process range. Historical data are used to build models to drive the output toward the solution with PID control providing the final drive to zero error.

The model is also used to develop dynamic tuning. The result is a simple to configure, robust solution to manage non-linear loop and/or disturbance variations.

(45w) Adsorption of CO2 by Alkaline Activated Ca-Al2O3-SiO2 Material

Naim Faqir, King Fahd University of Petroleum & Minerals, Dhahran, Saudi Arabia and Reyad Shawabkeh, Chemical Engineering, King Fahd University of Petroleum & Minerals, Dhahran, Saudi Arabia

Natural gas contains a slight level of carbon dioxide which leads to several industrial and domestic heating problems. In addition, it is one of main greenhouse gases that contribute for global warming, where its pollution is a wide-spread problem arising from a wide range of industries, leading to complex contaminant in atmosphere. Given the increasing pressures worldwide to minimize the emission of greenhouse gases into atmosphere, CO_2 must be remediated efficiently and cost effectively.

In this short abstract a chemically treated limestone and alumino-silicate material were utilized for removal of CO_2 from natural gas using a combined adsorption-chemical reaction processes. This functionalized Ca-Al_2O_3-SiO_2 material was characterized according to its crystalline phase, surface area, pore size distribution and surface morphology. Then it was tested against different concentrations of CO_2 in CO_2-N_2 gas streams.

Pressure Gauges: Hidden Danger or Visible Safety

by Jeff Placek, WIKA Instrument, LP, Lawrenceville, GA

Introduction

In recent years, refining plants, petrochemical plants and chemical plants have been required to increase capacity, but have been provided with very few upgrades to existing equipment to accommodate the increased capacity requirements. This has caused plants to have increased stress and strain on their systems and operating units. For example, higher pulsating pressures and heavy mechanically-induced vibrations are translating to all areas of the processing equipment, especially pressure instrumentation.

Pressure gauges are one of the most important, yet most overlooked or ignored keys to ensuring the continued safety, productivity and profitability of process plants. Process pressure gauges are used to read pressure on transfer lines, discharge pumps, and other places where it is important to be able to read pressures.

Problems Associated with Misapplied, Faulty or Ignored Gauges

Incorrect, misapplied, damaged, faulty or ignored gauges can lead to a number of issues, including safety problems, downtime, and costly regulatory agency fines and other penalties.

Process Safety:

One group of problems relates to process safety issues, such as loss of containment that leads to fugitive emissions, fires and explosions. Bourdon tubes in malfunctioning gauges can rupture, causing the release of media, such as gasoline, oil or gas. Because this media is often volatile and corrosive, it can cause fires, explosions, or other serious safety issues, such as a toxic gas releases or acids spraying onto workers.

Media releases are common results of misapplied, faulty or ignored gauges. Fires and explosions are less common but they do occur. When reviewing plant instrumentation, findings have revealed instances where the wrong gauges have been installed in locations on pumps. In one instance, a gauge vibrated so severely when the pump came out of alignment that it caused the Bourdon tube in the gauge to crack. The media coming out of the tube leaked onto the pump, which was excessively hot because of age and misalignment, causing a fire.

Expensive Downtime:

A second group of problems relates to expensive downtime, which can result from the breakdown of process safety just mentioned, or simply from having to shut down a process in order to replace incorrect, misapplied, broken or otherwise damaged gauges.

1 - Safety-Related Downtime:

As a result of the problems identified above (emissions, fires, explosions, etc.), there can be any number of causes of subsequent downtime, including fire suppression activities, released media containment, line shutdown, personnel evacuation, and then eventually having to rebuild or repair the lines, replace equipment, etc. In addition, if there is a media release with no injuries or fatalities, the plant is required to notify the EPA. However, if there is a media release that does cause injuries or deaths, the plant is required to notify OSHA. These can lead to very long process shutdowns, anywhere from a day to a month or more. (More on OSHA and the EPA below.)

2 - Non-Safety-Related Downtime:

When a line fails, causing a gauge to fail, even if there is no release or safety incident, downtime can still occur as a result of the process needing to be shut down in order to replace the gauge. Gauge failure rarely causes other parts of the process to fail, although gauge failure is usually an alert that there is a problem. In most cases, a plant worker will simply replace the pressure gauge and move on. When another failure occurs, the worker will do the same, until the root cause of the failure, such as a misaligned pump, causes the plant to shut down the line, resulting in more costly, unplanned downtime and a major maintenance time.

Regulatory Agency Fines and Penalties:

A third group of problems relates to the potential for significant fines and other penalties from federal regulatory agencies (primarily OSHA and the EPA), as well as from state and local jurisdictions, that result from compromised process safety.

1 - OSHA:

OSHA addresses many process-related safety issues in 1910 Subpart H, "Hazardous Materials." More specific is 1910.119, "Process safety management of highly hazardous chemicals." In its Process Safety Management section, OSHA notes that, "Unexpected releases of toxic, reactive, or flammable liquids and gases in processes involving highly hazardous chemicals have been reported for many years in various industries that use chemicals with such properties. Regardless of the industry that uses these hazardous chemicals, there is a potential for an accidental release at times they are not properly controlled, creating the possibility of disaster."

OSHA goes on to note that: "To help ensure safe and healthful workforces, OSHA has issued the Process Safety Management of Highly Hazardous Chemicals standard (29 CFR 1910.119), which contains requirements for the management of hazards associated with processes using highly hazardous chemicals."

OSHA's standard emphasizes the management of hazards associated with highly hazardous chemicals and establishes a comprehensive management program that integrates technologies, procedures, and management practices.

While OSHA does not monitor gauges in specifically, they will enforce guidelines for worker safety. OSHA investigates injuries and deaths, and may levy fines or other penalties.

2 - EPA:

According to Greg Neneman, director of business development solutions for Valin Corporation (San Jose, Calif.), fugitive emission regulations from the Federal level have tightened up over the past several years. In many states, such as California, Oregon and Washington, the regulations are even tighter.

"One challenge companies face is ensuring compliance with emission regulations when they upgrade or expand their processes," points out Neneman. "They are not allowed to produce additional emissions beyond the original limit set by regulations."

From the standpoint of federal EPA regulations, each company that produces or emits volatile organic compounds (VOCs) is required to come up with a leak detection and repair program. "In some areas, county and local requirements are even tighter than federal requirements," continues Neneman.

In certain areas, companies that produce or emit VOCs are required by regulation to do audits on valves, flanges, gauges, and all other process connections on a quarterly basis. "And in some jurisdictions, the regulatory agencies can give plants 24 hours notice before they descend on that plant to check all of their records and conduct other investigations," adds Neneman.

In May 2012, the U.S. Environmental Protection Agency (EPA) levied a $1 million civil penalty on a plastics company for alleged violations of the Clean Air Act at two of its chemical manufacturing plants. In addition to paying the penalty, the company agreed to implement a comprehensive program to reduce fugitive emissions of hazardous air pollutants (HAPs) from leaking equipment such as valves and pumps. The settlement also required the company to implement enhanced work practices, including more frequent leak monitoring, better repair practices, and new efforts designed to prevent leaks. After being fined by the EPA, the company realized that instrumentation was one of the areas where they having fugitive emissions. Until that time, workers had been replacing gauges left and right, because they weren't aware of the problems that could occur. The gauges weren't a big expense to them, so they considered the replacement activity to be more of a nuisance than anything. It's when they had the rupture that they realized there was a real problem.

An audit of the plant, which involved reviewing at all of the instrumentation, found that in many applications, they were using the wrong instrumentation. To correct the issues, the right instrumentation was specified for them, and the correct gauges were ordered. The plant now has its instrumentation under control.

While conforming to EPA regulations has always been a challenge for process facilities, it is going to become even more challenging in the future as a result of EPA's recent introduction of new requirements related to fugitive emissions and leak detection and repair (LDAR) strategies. On April 17, 2012, the EPA Administrator signed a suite of final air quality regulations that affect both upstream and midstream oil and gas sectors - the production/processing and transmission/storage sectors of the oil and natural gas industry.

For the first time, EPA will regulate air emissions from natural gas wells that are hydraulically fractured, as well as other emission sources associated with exploration, production, processing, and transportation of oil and natural gas.

Promulgation of the final rules will result in new control equipment requirements, as well as increased requirements for monitoring, recordkeeping, and reporting for owners, operators, contractors and vendors.

The final rules require the use of emission controls, as well as work practices, through two different air regulatory programs. One is the New Source Performance Standards (NSPS). The other is the National Emissions Standards for Hazardous Air Pollutants (NESHAP).

EPA finalized a new NSPS at 40 CFR Part 60, Subpart OOOO, which regulates emissions of volatile organic compounds (VOC) and sulfur dioxide (SO2) from a variety of oil and gas exploration, production, processing, and transportation facilities, including natural gas wells, certain compressors and pneumatic controllers, storage vessels, and sweetening units at onshore natural gas processing plants. Subpart OOOO also includes startup, shutdown and malfunction provisions.

The EPA has also made it clear that oil and gas production and processing sources will be the focus of its National Enforcement Initiative (NEI) for fiscal years 2012 and 2013. With the new NSPS and revised NESHAPs in place, EPA now has the tools it needs to enforce in a comprehensive manner.

As a result of these new regulations, plants should look at everything more closely, including gauges.

How Problems Can Occur With Gauges

Technical problems can occur with process pressure gauges in a number of ways, which can be conveniently categorized into seven broad categories with specialties or subsets. The most common reason for gauge problems is that workers often install gauges without knowing how they should be installed or accessorized. As a result, the gauges can fail because they were never designed to function in those specific situations. They can also fail because they were installed incorrectly, or because they were not properly maintained.

One way to eliminate or minimize these problems is to consider specific pressure gauge accessories. These can prove to be the extra step needed to ensure a successful, proactive preventive maintenance program, and for extending the life of the gauges. Accessories can provide additional protection for the gauges during their lifetimes.

1 - Mechanical Vibration.

This is oscillating, reciprocating or other periodic motion of a rigid or elastic body or medium forced from a position or state of equilibrium. Typically, vibration is the main cause of pressure gauge failure in process plants. Pump applications produce most of this vibration.

Gauges that aren't designed to withstand vibration may be installed in applications where vibration occurs, causing the gauges to come out of accuracy. Even more serious, a gauge can vibrate so much that the Bourdon tube inside it can crack, releasing media."

One way to prevent vibration-related problems is to install liquid-filled gauges, which help to dampen the vibration and also protect the internal components. The liquid case filling also lubricates the internals and prevents corrosion.

2 - Pulsation.

When media rapidly cycles through the pressure system of the gauge, pulsation occurs. If the pointer pulsation increments are greater than 5 percent of the full scale value, you must intervene to prevent damage to the gauge.

A restrictor or throttle screw can solve most pulsation problems. It will slow down the media by reducing the size of the orifice and averaging out pressure fluctuations. For extreme pulsation, such as spiking and surging, a snubber works best. It allows more control over the amount of snubbing and the flow of media into the pressure system.

3 - Temperature:

If a gauge is exposed to a media with very high temperature, the metal alloy in the gauge may not be able to handle the temperature, causing the Bourdon tube to fail and release media.

Long pipes that separate the pressure gauge from the process, or small cooling elements with radiator fins, protect the pressure instrument from high process temperatures by allowing ambient air to flow over the surface area. In environments with steam, using a siphon with the pressure gauge cools the steam down and drives it back to its liquid state before it gets to the gauge. With very high-temperature media, a diaphragm seal with a cooling element will isolate and prevent the hot media from reaching the gauge.

4 - Spikes or Overpressure.

In some cases, process media may be under extreme pressure. Pumps also can cause spikes and over-pressure, both of which shorten gauge life. Gauges can fail as a result of these pressure spikes. When workers turn pumps on or off, or open or close valves, a surge of media or oil goes through the pipe and hits the pressure gauge, which causes a huge spike. For example, if a line is pumping 100 psi of oil, someone may install a 100 psi gauge, not realizing that this leaves no allowance for any overpressure, surging, or spiking. When pressure does increase, it causes overpressure on the gauge, sometimes to the point of where the Bourdon tube ruptures and leaks media.

One way to prevent this problem is to use a different type of gauge, such as 200 psi gauge for a 100 psi line. Another option is to accessorize the gauge with an over-pressure protector, which would shut off the pressure if there were a surge or spike. Overpressure protectors are designed to avert spikes and the resulting overpressure damage. These protectors provide protection for pressure gauges against the effects of pressures exceeding their maximum pressure ratings.

5 - Corrosion.

Corrosion problems can occur with gauges if the process media is not compatible with the pressure gauges' wetted parts material. Media can sometimes be so corrosive that it corrodes through the Bourdon tube and leaks onto the ground.

When the media is not compatible with the wetted parts material of the pressure gauge, a diaphragm seal is required to isolate the gauge from the media. The benefit of a diaphragm seal is that it can be made of special alloys that are designed to withstand corrosion. It is essential to select a diaphragm material that is compatible to the process media.

6 - Clogging.

In some processes, the media may eventually crystallize or may otherwise harden. If a standard gauge is used in such an application, it will eventually get clogged and will not be able to read the pressure accurately. For example, if a gauge gets clogged and ends up reading zero pressure, when in fact it is under tremendous pressure, there will be an unexpected and dangerous release of media if and when the piping system is opened. Diaphragm seals protect the pressure gauge from solids, particles and viscous media by creating a barrier between the process and the pressure inlet of the gauge. There is also an option to equip the diaphragm seal with flushing ports to constantly flush material out from the diaphragm surface.

7 - Mishandling and Misapplication.

Gauges can also fail prematurely if they are misapplied. For example, a gauge may have been ideal for the application when it was installed. However, if there are equipment modifications or more components subsequently added, this can change process parameters, resulting in the gauge not longer being right for that application. Misapplication can lead to serious problems associated with failure and subsequent media release.

It is also possible to accidentally damage gauges during installation. Gauges should be installed using the proper sized wrench on the flats. Gauges should never be installed by turning the case. After installation, gauges can also fail if they are mishandled or used improperly.

Why Problems Can Occur With Gauges

In a typical process plant, about 25 percent of pressure gauges have failed or are about to fail.[1]
A lack of management focus and worker education and experience are key reasons plants experience issues with gauges.

Management Focus.

Gauges are often viewed as inexpensive commodities, with management failing to realize that they are very critical parts of the processes. Often, plants will replace failed gauges with the same replacement gauges, which will also fail, only to have to be replaced again. Eventually, their luck may run out and a serious failure may occur because the incorrect gauge has been installed.

Why do gauges tend to receive such little attention from management? Management tends to focus on the expensive equipment - $100,000 or million dollars items. They often don't understand the importance of focusing attention on $100 gauges. This is not the fault of management. The problem is that no one has explained to them just how important gauges are.

As noted earlier, though, ignoring instrumentation can cause serious safety and financial problems. Plant process gauges can be thought of in the same way as vehicle gauges. If the temperature gauge on an automobile is defective, it is possible to destroy the engine as a result of overheating that is not registered by the defective temperature gauge. Similarly, if the oil pressure gauge on an automobile is defective, it is again possible to destroy the engine as a result of seized components resulting from loss of oil that is not indicated by the defective oil pressure gauge.

Worker Education and Experience.

A lack of worker education and experience also contributes to gauge issues. In many instances, when maintenance technicians look at gauges, they don't pay much attention to them unless they leaking or look as if they are going to explode. They don't realize all of the information that that gauge can provide. In addition, technicians may not realize that a gauge has failed or is about to fail. Again, as long as it is not leaking media, they will often just keep walking. This is rarely the fault of the technicians. Many have not been properly trained to understand process pressure gauge performance.

Along the same lines, the industry has been experiencing a "brain drain" for several years, with large numbers of experienced technical workers and engineers retiring and being replaced by younger people who lack the education and experience to understand what gauges are telling them, or, worse, not being replaced at all.

According to research conducted by Cambridge Energy Research Associates in 2007, more than half of all oil-field professionals will reach retirement age during the next decade. The Interstate Oil & Gas Compact Commission found that the average age of production and service company employees is 46 to 49 years, and the average retirement age in the oil and gas industry is 55 years.[2]

Historically, plants had specialists whose jobs focused exclusively on instrumentation. They spec'd the gauges, maintained them, and made sure they were operating correctly. However, as plants got leaner in recent years, they started letting these people go. Plants first shifted gauge responsibility to the pipefitter whose main responsibility was to make sure the pipes were put together correctly and not leaking.

[1]WIKA data aggregated from more than 250 instrument audits.
[2]Peter Peek, Jocelyn Fenard, Patrick Gantes, Christine Theiler. December 2008, "Skills Shortage in the Global Oil and Gas Industry: How to close the gap" Part I. Pages 14-15.

Then, as plants began letting pipefitters go, they shifted the responsibility to the unit operators or technicians, who now install the gauges. These people usually don't have any experience or knowledge of how to install instrumentation. They usually just replace old gauges with similar gauges. The problem is that the gauges are usually the wrong ones to begin with. The result is that, these days, few plants have people who have much if any education and experience with instrumentation.

Taking Stock of Process Pressure Gauges

When it comes to improving the safety, productivity and profitability of process plants through an effective gauge management program, companies can take a couple of steps. First, an understanding of the vital importance of gauges must be established. Second, workers must understand what kind of gauges will be effective.

Focus on Gauges:

World-class process plants maintain their assets proactively. In such plants, pressure gauges receive as much attention as pumps and control valves. Management understands the importance of proactively maintaining their assets, and they realize that one of the best ways to maintain their most expensive assets is to properly source and maintain the gauges that protect these assets. That is, they understand the value of spending $100 for a gauge that can subsequently prevent damage or destruction to a $100,000 or $500,000 piece of equipment and the resulting negative safety, downtime and regulatory implications.

There is an interest in predictive maintenance these days in process plants, which is a good thing. A lot of this, though, involves expensive software. In addition, when something is out of kilter, the system may be set up such that technicians may have to wait for it to get far enough out of the normal range before it will generate an alert on the computer. However, properly-maintained pressure gauges, studied by people who know what to look for, can be very inexpensive but reliable and useful mechanical predictive maintenance tools.

Gauge Procurement Considerations:

Over time, as noted earlier, the status of pressure gauges has unfortunately diminished to that of commodities in many process plants. In order to ensure process plant safety, productivity and profitability, it is essential that plant professionals understand the importance of selecting the right gauges for applications.

It is also important that these be high-quality gauges, not "commodity" gauges. That is, not all process gauges are created equal. The so-called "lowest price" gauges appear attractive at first, but are rarely the "lowest cost" in the long run, because they usually have to be replaced more often, leading to excessive downtime and inconvenience. Worse, they may fail unexpectedly, leading to catastrophic results.

Consider these 10 steps to select the right gauges that will help to ensure safe, productive and profitable plant processes.

1 - Know what is in the pipe. This way, you can select a gauge alloy that will be compatible with the process media.

2 - Next, know the temperature of the media. If it is very hot, for example, it will be unsafe to use a standard pressure gauge. A long pipe, siphon or diaphragm seal assembly may be needed to cool the media sufficiently before it reaches the gauge internals.

3 – Third, know the ambient conditions surrounding the gauge. For example, if it is corrosive around the gauge, you need to make sure that the external components of the gauge such as the case and window can handle the environmental conditions.

4 - Next, know the pressure in the piping that the gauge is meant to measure. A best practice is using a gauge that is designed for two times the normal operating pressure. For example, if normal pressure is 100 psi, use a 200 psi gauge.

5 – Know the conditions that affect gauge wear. These could include mechanical vibration, pulsation, pressure spikes, etc. For example, if vibration exists, a gauge with a liquid-filled case is recommended. If there is pulsation, use a throttle plug or restrictor accessory.

6 - Select gauges that are designed for a particular service. Process type gauges are designed with strong Bourdon tubes and other heavy duty components. If a gauge is misapplied, the internals of the gauge may not be rugged enough to resist failure and subsequent media release in these applications.

7 - The gauge case must be rugged and have the correct safety features. In some case designs, if there is a rupture of the Bourdon tube, media will be released out the front of the gauge, which can spray workers. Select a gauge where media would release out from the back of the gauge.

8 - Accuracy is important. In most process plants, +/-0.5% accuracy is the norm, and +/-1% is acceptable.

9 - Size is also important. The gauge dial needs to be large enough that operators can read it from a distance – four inches or larger.

10 - Finally, the gauge should be mounted properly. A gauge is usually mounted by its stem, but it can also be mounted in a panel or against a flat surface.

Looking Outside for Assistance

As noted earlier, it can be difficult for plant management to devote enough time and attention to an effective process pressure gauge program. Similarly, it can be difficult for technical personnel to get up-to-speed in terms of education and experience related to process pressure gauges.

There has been a growing trend over the last 30 years in virtually all industries for companies to outsource services that are not considered "core competencies." That is, companies have come to realize that it makes sense to focus their attention on their missions and what they do best, and leave secondary and peripheral activities to third parties that specialize in these other activities and services. Becoming experts in pressure gauges is certainly not a core competency for refining plants, petrochemical, and chemical plants, so this is an area that can easily be outsourced to an experienced third-party audit team.

If a plant elects not to gain the experience that it needs on its own, free assistance is available. To ensure that gauges are working properly and are able to protect the valuable assets that they were designed to protect, a five-phase instrument review is recommended. A best practices review identifies gauges with unwanted conditions and recommends corrective actions to ensure optimum performance. The review identifies issues that affect reliability, longevity and safety. Such a review can uncover evidence of pulsation, vibration, pressure spikes, excess process or ambient temperature, installation damage, clogging, chemical incompatibility, and other problems that can lead to serious problems.

Phase 1 - Interview. This involves discussions between the third-party audit team and plant personnel. During the interview, plant personnel learn the purpose of the instrument audit, how it will be conducted, and how they will benefit from the audit. The interview also helps determine the best way to match the audit to the needs of that specific plant.

Phase 2 - Pre-Meeting. This involves determining the units or sections of the plant to audit, as well as outline roles and responsibilities. It also identifies prohibited and exempt areas of the plant. The third-party team reviews problematic instrument installations and discusses safety procedures with plant personnel.

Phase 3 - Instrument Application Audit. This involves having key plant personnel lead the third-party team on an actual plant walk-through. The audit team reviews the target gauge installations, identifies gauges that demonstrate unwanted conditions, as well as those that demonstrate installation best practices.

Phase 4 - Post-Meeting. This involves explaining the instrument audit results to key plant personnel, specifically noting gauge installations that pose immediate safety concerns. Together, the audit team and plant personnel determine which installations to investigate immediately and which to block out. The audit team and plant personnel then set schedules for the follow-up meeting.

Phase 5 - Follow-up Meeting and Presentation. The third-party team presents its general findings and the executive summary resulting from the instrument audit. The team also presents a spreadsheet that lists details about the gauges from the instrument review, color-coded by severity. The team also provides recommendations for improvements based on experience and best practices. Together, the audit team and plant personnel develop a strategy for moving forward.

The plant should also consider leveraging the audit team in other ways. For example, over time a plant's storeroom inventory can get out of hand. A qualified third-party audit team can help plant personnel rationalize, consolidate and manage their inventory of pressure gauges. For example, if a plant stocks spares for 10 or more ranges, the audit team may be able to help reduce that number to five or six ranges.

Essentially, when used properly, gauges play a significant role in plant safety, productivity and profitability. They act like silent alarms, indicating when something is seriously wrong or helping to diagnose deeper issues within systems and processes that can eventually lead to serious safety issues. To ensure safety and reliability at your plant, make certain the proper gauges are in place by reviewing your processes, gauges and your inventory. Everything is at stake.

REFERENCES

1. WIKA data aggregated from more than 250 instrument audits.

2. Peter Peek, Jocelyn Fenard, Patrick Gantes, Christine Theiler. December 2008, "Skills Shortage in the Global Oil and Gas Industry: How to close the gap" Part I. Pages 14-15.

ExxonMobil
Research and Engineering

Best Practices in Pilot Plant Piping

Richard Palluzi
April, 2013

The ideal piping for a pilot plant would be:

- Leak resistant
- Easily modified
- Able to accommodate varying requirements
- Easy to install
- Low cost
- Available in many different materials of construction
- Available in all sizes
- Have fittings available in numerous configurations

Leak Resistance

- Pilot Plants are more subject to leakage than process plants
  - Smaller size
  - Larger number of joints
  - Constant modifications
  - Need to accommodate accessibility, flexibility and change
- Common types of piping used in pilot plants
  - Piping
    + Threaded
    + Welded and flanged
    + Groved
  - Tubing
    + Compression fittings
    + Vacuum fittings
    + Mechanically joined

Best Practice: Tubing

- Available in numerous materials
- Fittings are readily available up to 1" OD
- Easier to install and modify
- Requires less skill, tools and training for installation
- Smaller and lighter than pipe
- Lower cost than pipe
- Ability to bend
  - Reduces need for fittings (lower cost and less potential for leakage)
  - Easier to install in tighter locations
  - Allows for easier removal of components for cleaning, modification or inspection

Best Practice: Compression Fittings

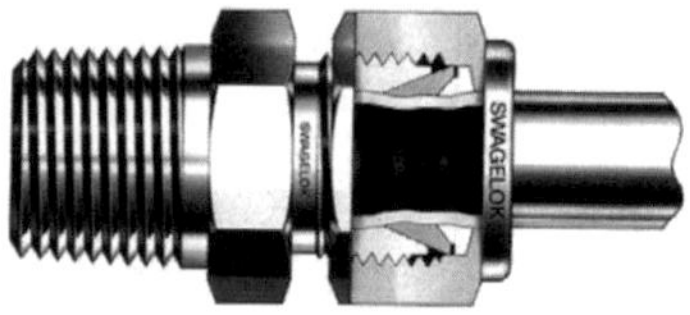

- Better leak resistance than threaded fittings
- Requires few tools and less training
- More expensive than pipe fittings
  - Significantly faster to install so labor savings usually make then cheaper overall
- Each fitting is a potential disassembly point
  - Do require clearance although much less than piping

- Significantly better make/break life than pipe fittings
  - But not infinite!

- Numerous specialty fittings available
- Many small valves and associated components available
- Most compression fittings have limitations on tubing hardness
  - Confirm the tubing you ordered is within specification

5

ExxonMobil
Research and Engineering

Best Practice: Compression Fittings

- Recognize that compression fittings larger than ¾" may require mechanical assistance
- Compression fittings have both minimum and maximum wall thicknesses
  - Too thin and the tube may collapse without swaging
  - Too thick and the tube will not deform enough to prevent blowout
- Drilled through compression fittings, used for dip tubes, thermocouples and similar uses, lose significant pressure ratings

OD(inches)	Typical Pressure Derating
½ and less	25%
5/8-3/4	50%
7/8 - 1	75%

6

ExxonMobil
Research and Engineering

Best Practice: Vacuum Fittings

- Significantly better leak performance than threaded fittings
  - Usually requires some welding to avoid additional joints
- More expensive than compression and pipe fittings
- Each fitting is a potential disassembly point
- Zero clearance fittings
- Long term make/break life
- Gasket surface is more prone to damage
  - Awareness can reduce frequency
- Decision to use O-ring style or metal gasket is usually based on temperature and pressure

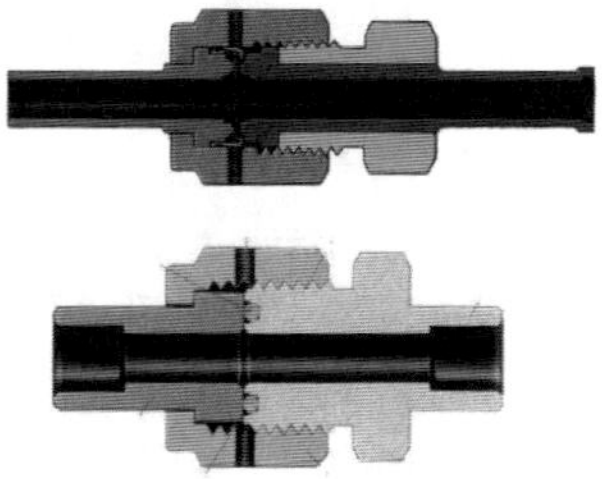

ExxonMobil
Research and Engineering

Best Practice: Threaded Piping

- Make sure the die is sharp and in good condition or the threads are likely to be rounded and more prone to leakage
- Make sure the threading is the right length
  - Longer or shorter lengths will leak
- Use good quality fittings
  - Particularly unions
- Avoid threaded connections in cyclic service
  - Temperature
  - Pressure
- Lightly coat unions with lubricating oil before each assembly
  - The coating greatly helps it to seal in gas service
- Recognize that threads over 1" NPS are increasingly difficult to seal
  - Particularly in gas service

ExxonMobil
Research and Engineering

Best Practices: Flanged Pipe

- Use raised face flanges whenever possible as they require less force to seal
- Avoid threaded flanges
  - Adds another potential leak point
  - Often makes bolt alignment difficult if retightened
- Tubing is usually a better approach for pipe sizes ½" and smaller

ExxonMobil
Research and Engineering

Best Practice: Specialty Piping

- Understand when specialty piping may be a viable or more effective option
- Grooved piping allows for fast and easy disassembly
  - Lower pressure (1,000 psig is typical)
  - CS or SS only
  - High materials costs
  - Some specialty equipment (grooving tool) required
  - Sizes below 1" NPS very limited

- Sanitary fittings allow for ease of cleaning
  - Available for piping or tubing
  - 304 or 316 SS only
  - Lower pressure (300 psig is typical)
  - High materials costs
  - Sizes below 1/2" NPS very limited

ExxonMobil
Research and Engineering

Best Practice: Specialty Piping

- Flare compression fittings are excellent for thermal cycling but are generally only available in CS and brass
  - Good for refrigeration systems
- Longer ferrule designs and cone and collar type compression fittings extend the range of compression fittings to 15,000 or 60,000 psig respectively
  - Much more expensive

11

ExxonMobil
Research and Engineering

Best Practice: Mounting

- Use vacuum fittings where frequent removal or cleaning is required
  - Zero clearance makes layout easier
  - Ability to replace gaskets maximizes service life

- Mount the valves not the tubing wherever possible
  - Minimizes the stress on the fittings created when operating the valves
  - Will reduce long term leakage
- Avoid unsupported elements

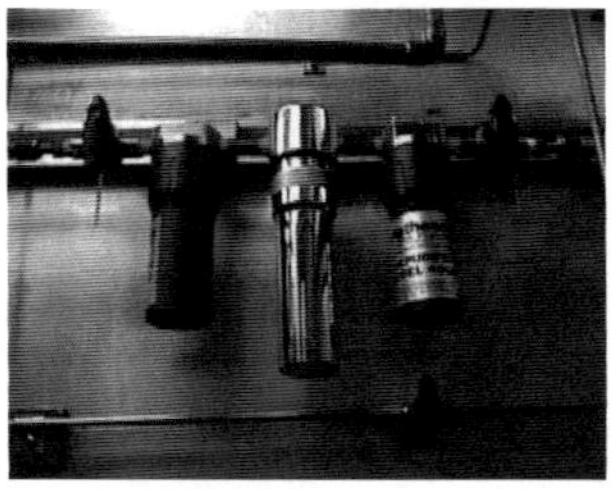

  - Long spans
  - Heavier components
  - Flimsy supports
  - All lead to increased leakage
- Use bends to allow for thermal expansion
  - Even small temperature swings promote leaks over time

12

ExxonMobil
Research and Engineering

Best Practice: Reduce Joints

- Fewer joints always mean fewer potential leak points
- Use specialty fittings to minimize joints
  - Long term savings outweigh costs over time
  - Requires some planning or stocking
- Weld joints that are:
  - In difficult to reach locations
  - Not likely to need to be opened
  - Are on systems subject to routine or frequent temperature swings
  - Under insulation

Best Practice: Remember Maintenance

- Almost everything eventually needs maintenance
- Make sure easy removal is provided for:
  - Instruments
  - Rotating equipment
  - Filters and strainers that need to be removed to be changed
  - Reactors
  - Regulators
  - Relief devices
  - Control valves
- Decide if you have clearance or need zero clearance fittings

Best Practice: Leave Room

- Pilot plants are always modified
- Building a unit in too small a space:
  - Adds to construction costs
  - Makes future changes difficult or impossible
  - Greatly increases maintenance costs
- The Golden Rule: Plan on at least twice the largest space you think you'll need
- Piping runs eat up space
- Insulation easts up space

ExxonMobil
Research and Engineering

Best Practice: Training

- Train your operators and craft support regarding best piping practices
- Always use a holdback whenever making or breaking any fitting
  - Otherwise the fittings o neither side will start to leak
- Make up the fitting the proper amount
  - Not too much, not too little
  - Consider gauge blocks for compression fittings
- Never remove a support
- Make sure any changes or modifications are properly supported

ExxonMobil
Research and Engineering

Best Practice: You Get What You Pay For

- Buy good quality pipe, tubing and fittings
- Use high quality gaskets and sealants
  - Teflon tape and pastes generally work better than dopes
  - Pastes require a minimum time for hardening
  - Silver plated vacuum gaskets seal better than plain stainless
  - Spiral re-enforced gaskets seal easier and last longer than flat gaskets
  - Paying more for a good quality seal always ends up being more cost effective!

17

ExxonMobil
Research and Engineering

Best Practice: Heat Tracing

- Avoid joints wherever possible
- Use bends to accommodate thermal expansion
- Use standoffs to avoid having supports becoming heat sinks
- Use vacuum fittings when joints are necessary
- Mark insulation for joint location for ease of maintenance

18

ExxonMobil
Research and Engineering

(54b) Successful Operation of a 1 BPD Fischer-Tropsch Pilot Plant

Juan Inga and **Lindsay Hernandez**, SGC Energia, Houston, TX

In 2012 SGC Energia commissioned, started up, and operated a pilot scale Fischer-Tropsch process for the demonstration step of the XTLH solution in Pasadena, TX. This plant was used to confirm design plans for future larger scale projects as well as to gather additional data. Running a Fischer-Tropsch reactor at a nominal capacity of 1 BPD, SGC Energia gained valuable information from lessons learned and the challenges of the start-up process. Upon the conclusion of the successful run, this moveable skid is being relocated to provide future operator training for a commercial unit. This commercial plant will be located in the Gulf area.

SGC Energia has operated different FT reactor technologies at a 1BPD level, plus others on a smaller scale. Commercial sized samples have been produced from these operations. Currently the SGCE Fischer-Tropsch reactor choice greatly reduces the risk involved, making the system more robust; and therefore more equipped for commercialization.

(54c) Continuous Improvement in Pilot Plants for Safer Operation

Kiran Gupte, BP Refining & Logistics Technology, Naperville, IL

Continuous Improvement in Pilot Plants

for Safer Operation

Kiran Gupte

BP Refining & Logistics Technology

150 W. Warrenville Rd

Naperville, IL 60563

BP's Refining & Logistics Technology group has successfully designed, commissioned and operated pilot plants in support of refining processes over the last 20 years. Of course, over the years the management practices have evolved with newer methodologies. The "Continuous Improvement Process" is important to critically review all operations to further strengthen the safety in design and operation. In addition to the old and reliable "Process Safety Review (PSR)" process, the newer "Management of Change (MOC)" procedures play the most critical role. An "Inherently Safer Design" is not only for a new pilot plant during the conceptual design phase but it should also be considered in every improvement. The "Continuous Improvement" process helps to bring out and consider smaller steps in improving operation of the pilot plants. The cumulative effect of every improvement step however can never be over emphasized. This presentation will highlight the challenges of operating small scale pilot plant and methods to continually improve these units for safer operation. As the pilot units get used for ever expanding types of projects, the attention to "Operating within the Design Envelope" is important for safe operation with zero incidents. All these aspects are an integral part of the overarching Operations Management System. The safe practices developed are applicable to all types of equipment including small-scale catalyst screening reactors, small-medium autoclave reactors, distillation apparatus, and large pilot plant and demonstration reactors handling a broad range of hydrocarbon feedstocks.

(54d) Using Dimensional Analysis to Design Pilot Plant Experimental Programs

Jonathan H. Worstell, Consultant, Richmond, TX

Pilot plants are capital intensive and expensive to operate. The most expensive item of a pilot plant budget is its experimental program. A poorly designed pilot plant experimental program requires more operating campaigns than originally budgeted. Thus, such a pilot plant will "go beyond" its budget --- a non-optimal situation.

This presentation discusses using Dimensional Analysis as a method for designing pilot plant experimental programs. First, Dimensional Analysis helps identify the major parameters impacting the chemical process. Second, Dimensional Analysis shows how to minimize the experimental program for a given pilot plant, thereby reducing its total operating expense.

Dimensional Analysis is not new: it has just fell out of favor with chemical engineers. It is time to rethink its use.

Design of Batch Scheduling

Aaron Herrick, Chemstations, Inc.

Chemstations™ www.chemstations.com

Abstract

Design and scale-up of new batch processes presents a unique set of challenges versus design of continuous processes. One key difference is optimization of the time required for each process step. Rules of thumb are often used due to lack of data or tools.

Improvements in dynamic simulation software have allowed for more rigorous modeling of increasingly complex systems and scenarios. Modeling of batch process steps in the design phase can highlight potential problem areas such as heat balance limitations, short residence times, reversal of flows, and pressure/flow dependencies. This allows for efficient optimization of step scheduling using the latest dynamic simulation techniques. Examples of batch designs are presented using CHEMCAD simulation software. A new method is included for simplification of handling any number of process steps and control elements.

Chemstations™ 1

Agenda

- Current practices in design of batch scheduling
- Improvements made possible with
  - Rigorous modeling methodology
  - Microsoft Excel™ to drive a schedule in a simulator
- Example using CITRON
- Example using WINTEK
- Summary & suggested procedure

Chemstations™ 2

Current practices in design of batch scheduling

- Methods
  - Best guess & experience
  - Lab scale & pilot scale testing
  - Use a simulation
- Pros
  - Fast
  - Fairly easy
  - Gives a basic time and heat duty analysis
- Cons
  - Suboptimal schedule times
  - Might not account for equipment/process limitations
  - Difficult to do safety analysis until after process is running

Chemstations™ 3

CITRON (current practice)

Recover 120 kgs of 99.9999% pure Citronellal from 170 kgs essential oil using a 500L pot still with a 10ft packed column.

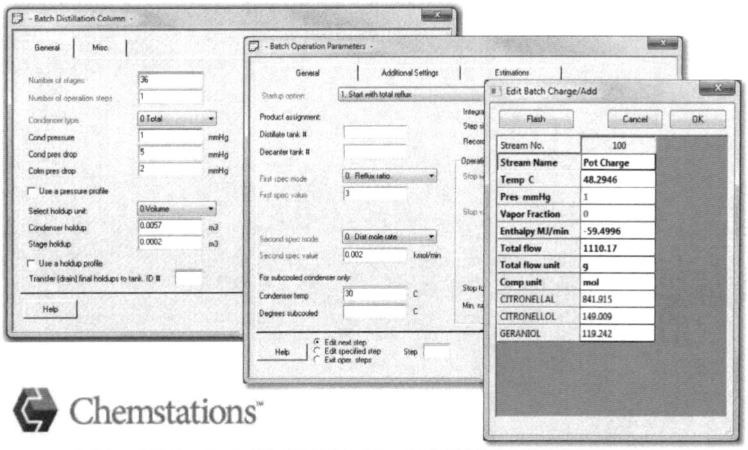

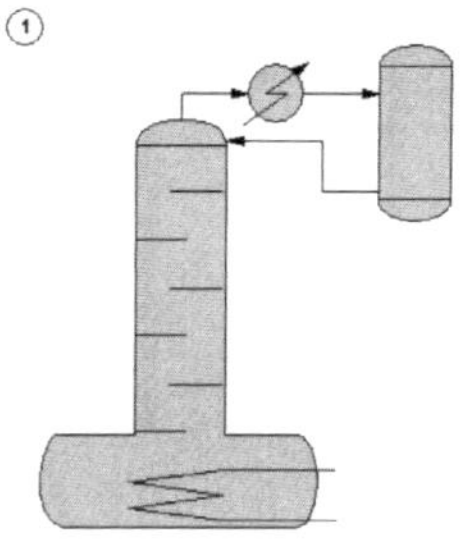

Chemstations™

4

CITRON (current practice)

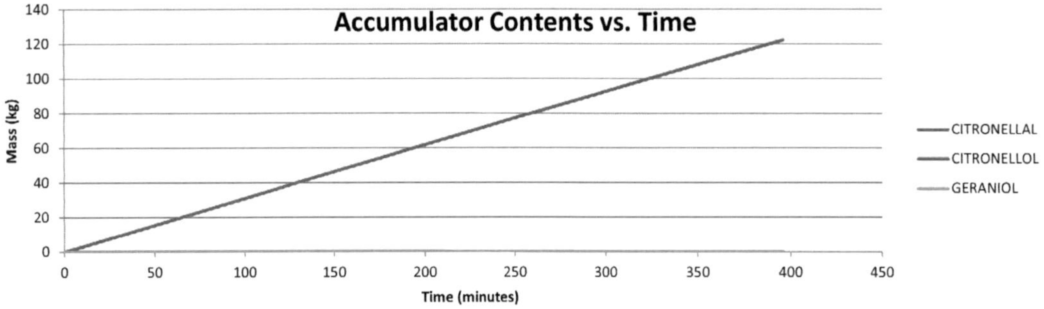

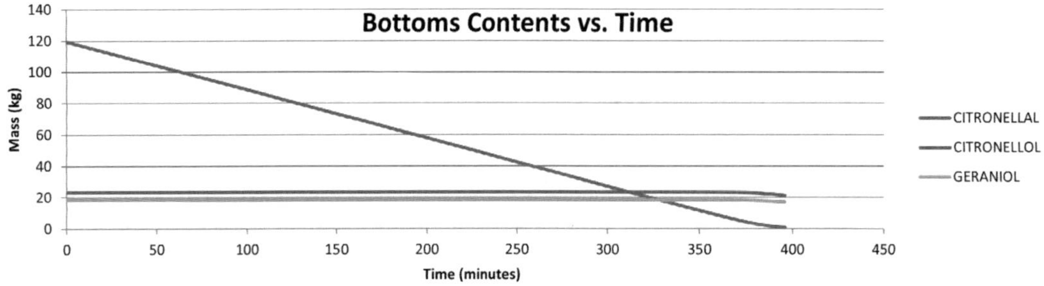

Chemstations™

5

CITRON (current practice)

What have we learned from the model?

- Heat duty/flow/time relationship
- Basic operation steps/timing
- Limited equipment sizes and specifications:
  - Heating requirements
  - Condenser requirements
  - Column dimensions

Chemstations™ 6

CITRON Dynamic (advanced practice)

- Expand our previous example to include
  - Rigorous heat exchanger geometry and performance
  - Event sequencing using DATAMAP to Microsoft Excel
  - Utilities modeling
  - Column metal heat transfer
  - Heat duty/ cooling water on control
  - Dry column startup to total reflux
  - Detailed engineering, e.g., nitrogen sweep on vessel, insulation, etc.

Chemstations™ 7

CITRON Dynamic

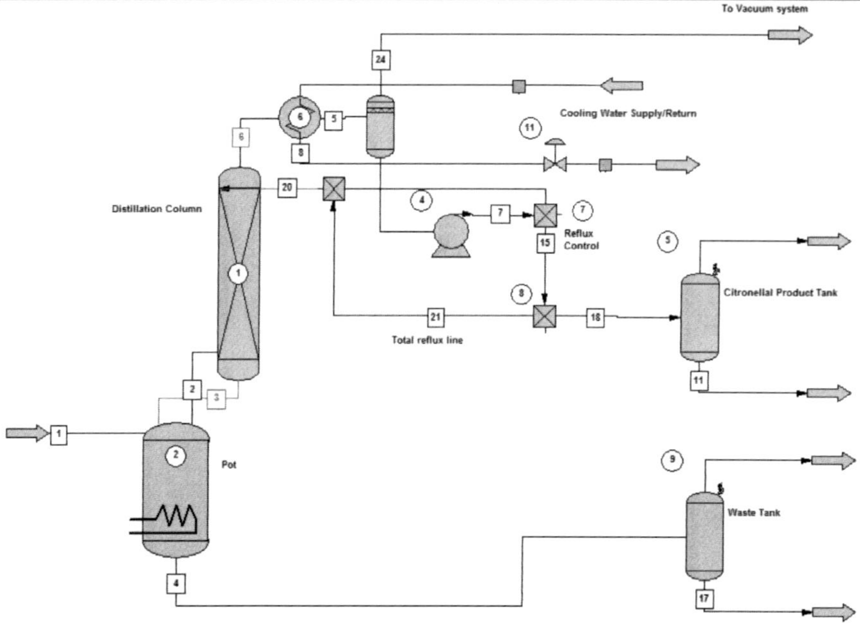

Chemstations™

Schedule of Events

- A schedule of events is made in Excel and must be connected to CHEMCAD:

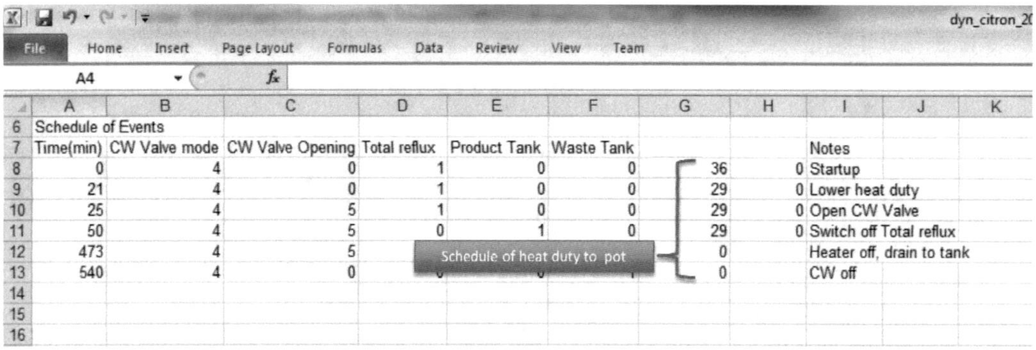

	A	B	C	D	E	F	G	H	I
6	Schedule of Events								
7	Time(min)	CW Valve mode	CW Valve Opening	Total reflux	Product Tank	Waste Tank			Notes
8	0	4	0	1	0	0	36	0	Startup
9	21	4	0	1	0	0	29	0	Lower heat duty
10	25	4	5	1	0	0	29	0	Open CW Valve
11	50	4	5	0	1	0	29	0	Switch off Total reflux
12	473	4	5	[illegible]	[illegible]	[illegible]	0		Heater off, drain to tank
13	540	4	0	0	0	1	0		CW off

Chemstations™

CHEMCAD to Excel connection

- CHEMCAD sends time information to Excel cell A2 via datamap
- Excel uses time, schedule, and VLOOKUP() function to determine values of process variables
- CHEMCAD collects the current time step data from row 2 of the Excel sheet and enters it into the flowsheet via datamap

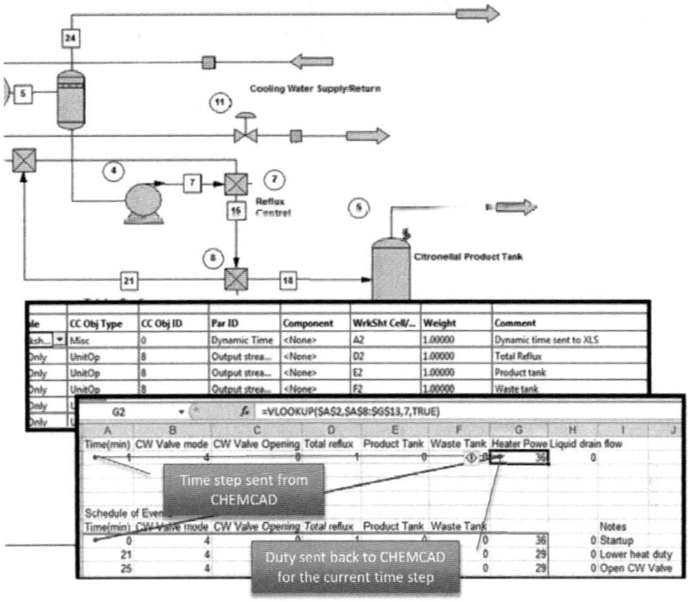

Chemstations™

10

CITRON Dynamic Operations

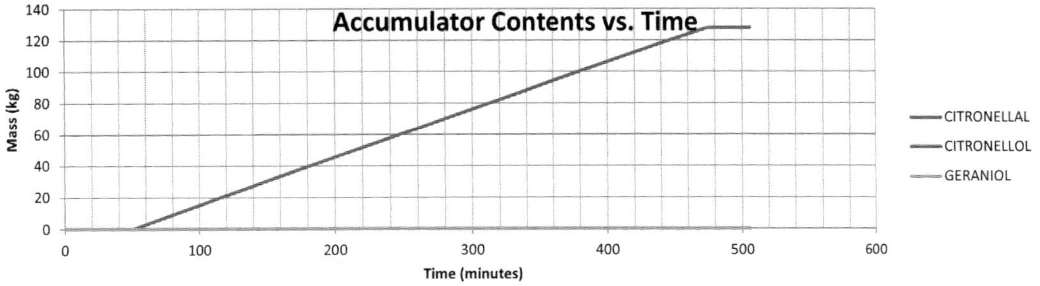

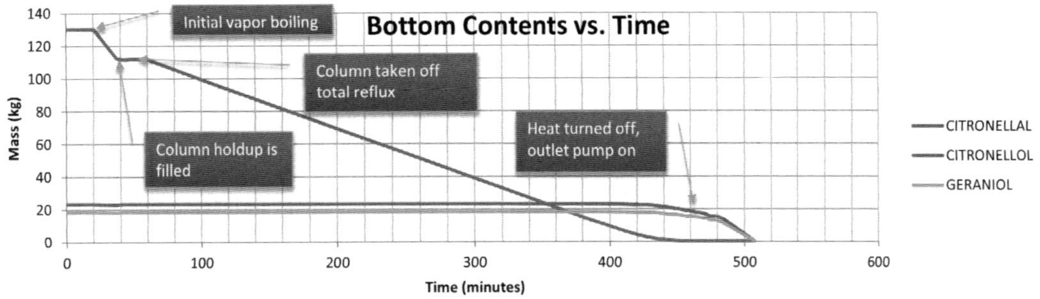

Chemstations™

11

Comparison of CC-BATCH and CC-Dynamics Column Bottom Flow

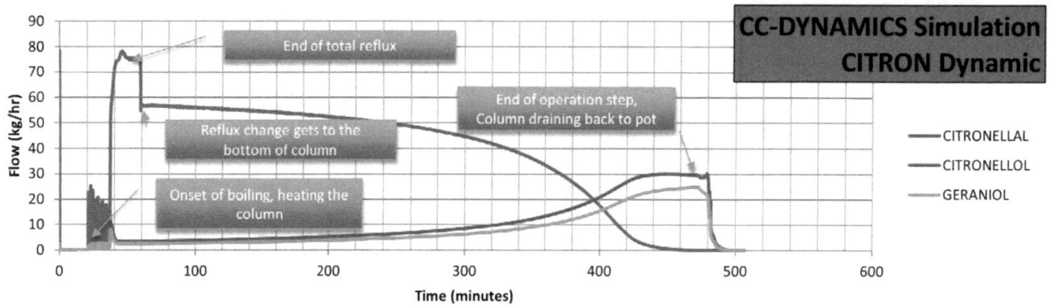

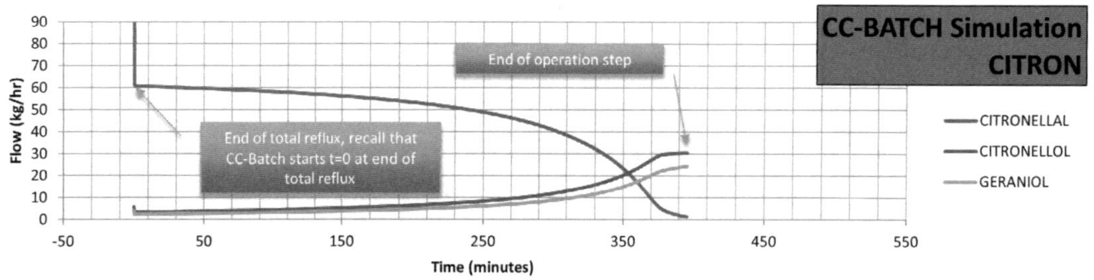

Chemstations™

12

CITRON Dynamic Conclusions

We learned from using dynamics:

- Detailed startup procedure
- Detailed operation steps/Sequence details
- Equipment performance limitations
- Vacuum load
- Utility demands
- Equipment optimization is now possible: checking condenser capacity, vacuum system capacity, column flood %, column insulation requirements, etc.
- Higher fidelity simulation provides higher fidelity economic calculation (campaign time and costs)

Chemstations™

13

WINTEK Batch Dehydration

Skid mounted solvent dehydration plant

Stripping water from a solvent stream in a two bed adsorber (mol sieve) system. One bed is active and one bed is regenerating under vacuum. Process was already built and operational before modeling analysis.

Process is scheduled with a defined sequence for opening and closing valves to allow one bed to regenerate while the other adsorbs. Rigorous equipment and piping models and pressure/flow calculation included (allows for reversible flow).

Rigorous simulation allows us to simulate effect of malfunction (RB3 blowdown valve malfunction)

Chemstations™ 14

WINTEK Process Flowsheet

2 Adsorption beds

2 PID controls

10 Actuated ball valves

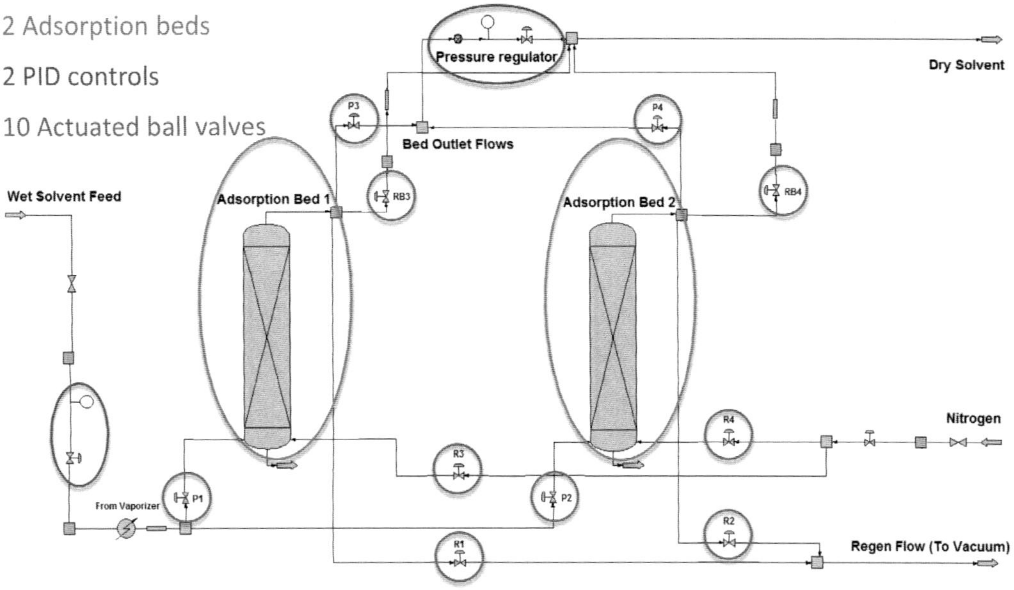

Chemstations™ 15

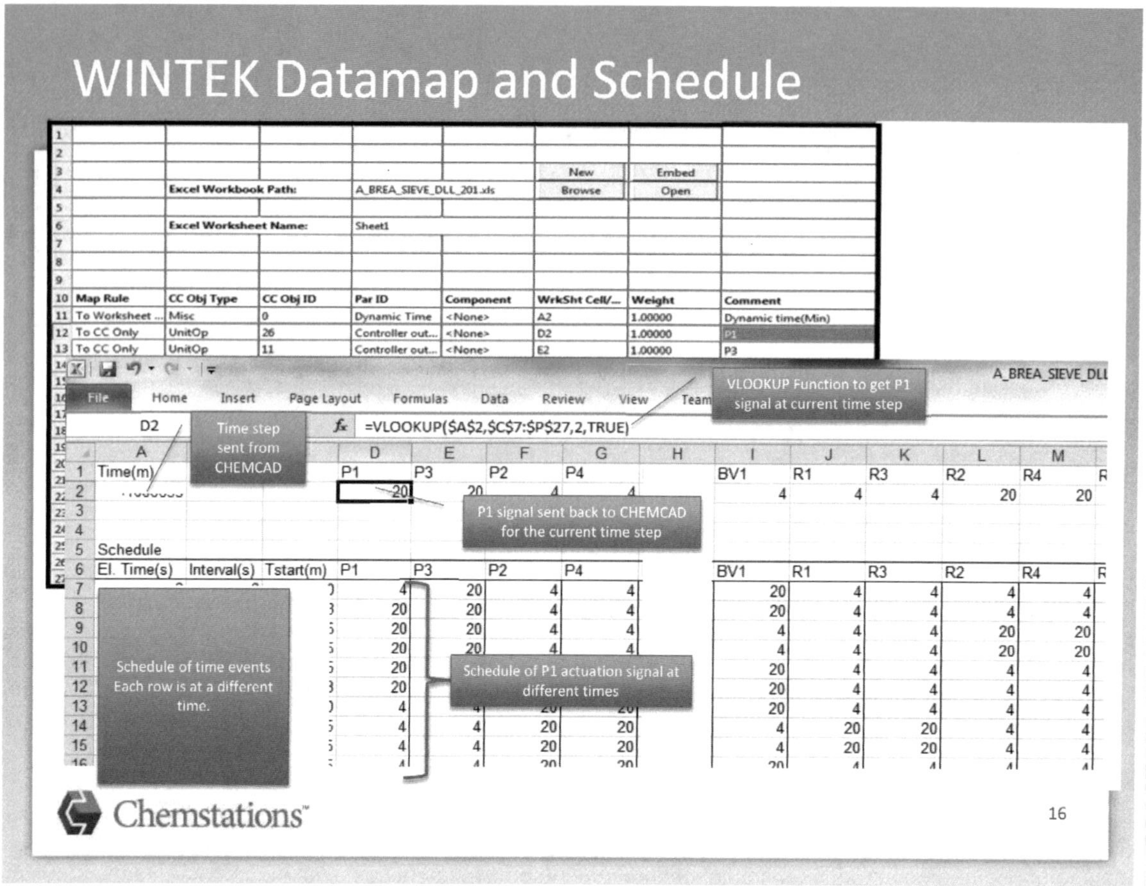
WINTEK Datamap and Schedule
Excel Workbook Path:
A_BREA_SIEVE_DLL_201.xls
New
Browse
Embed
Open
Excel Worksheet Name:
Sheet1
Map Rule
CC Obj Type
CC Obj ID
Par ID
Component
WrkSht Cell/...
Weight
Comment
To Worksheet ...
Misc
0
Dynamic Time
<None>
A2
1.00000
Dynamic time(Min)
To CC Only
UnitOp
26
Controller out...
<None>
D2
1.00000
P1
To CC Only
UnitOp
11
Controller out...
<None>
E2
1.00000
P3
File
Home
Insert
Page Layout
Formulas
Data
Review
View
Team
A_BREA_SIEVE_DLL
D2
=VLOOKUP($A$2,$C$7:$P$27,2,TRUE)
VLOOKUP Function to get P1 signal at current time step
Time step sent from CHEMCAD
P1 signal sent back to CHEMCAD for the current time step
Time(m)
Schedule
El. Time(s)
Interval(s)
Tstart(m)
P1
P3
P2
P4
BV1
R1
R3
R2
R4
Schedule of time events Each row is at a different time.
Schedule of P1 actuation signal at different times
Chemstations
16

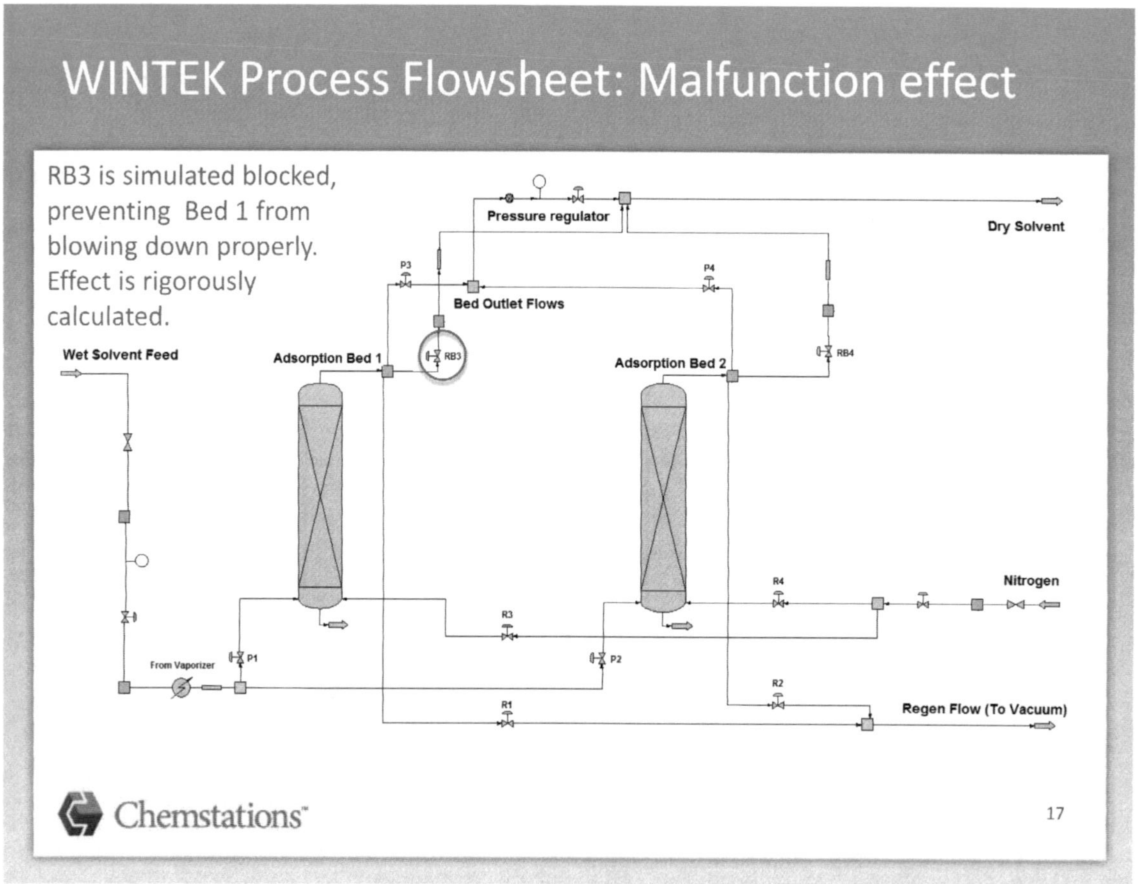
WINTEK Process Flowsheet: Malfunction effect
RB3 is simulated blocked, preventing Bed 1 from blowing down properly. Effect is rigorously calculated.
Pressure regulator
Dry Solvent
P3
P4
Bed Outlet Flows
Wet Solvent Feed
Adsorption Bed 1
RB3
Adsorption Bed 2
RB4
R4
Nitrogen
R3
From Vaporizer
P1
P2
R2
R1
Regen Flow (To Vacuum)
Chemstations
17

WINTEK Adsorber Beds

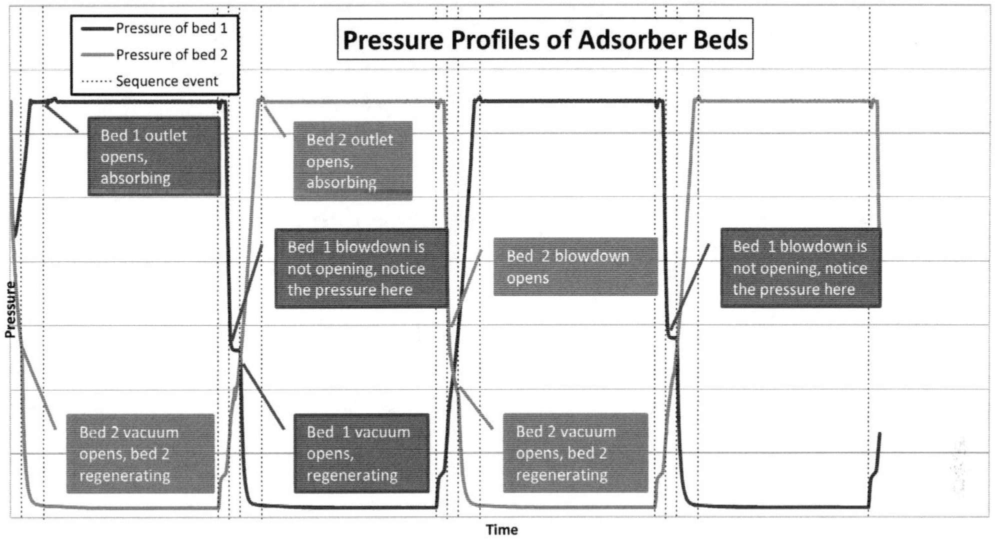

Chemstations™ 18

WINTEK Anhydrous Solvent Flowrate

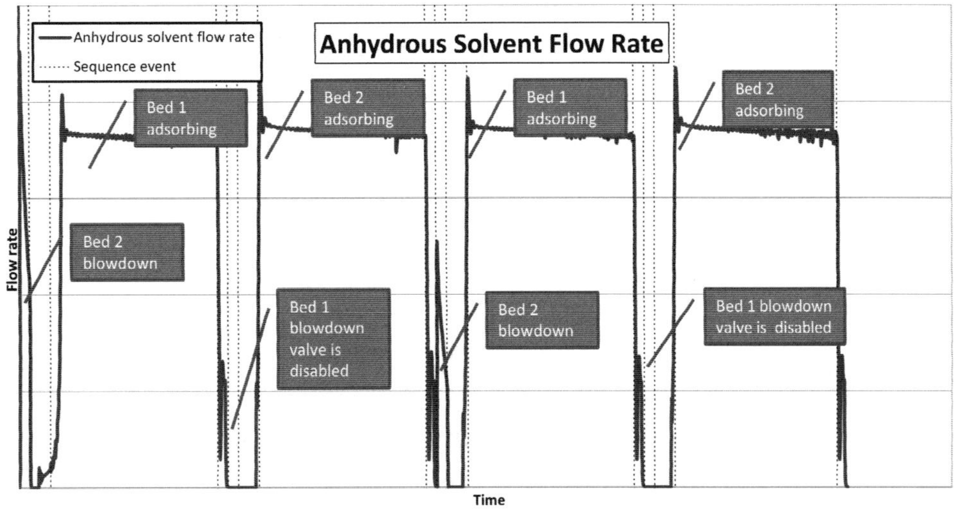

Chemstations™ 19

WINTEK Process Flowsheet: reversing flow

P3 and P4 open, flow is forward through one valve, reverse through the other

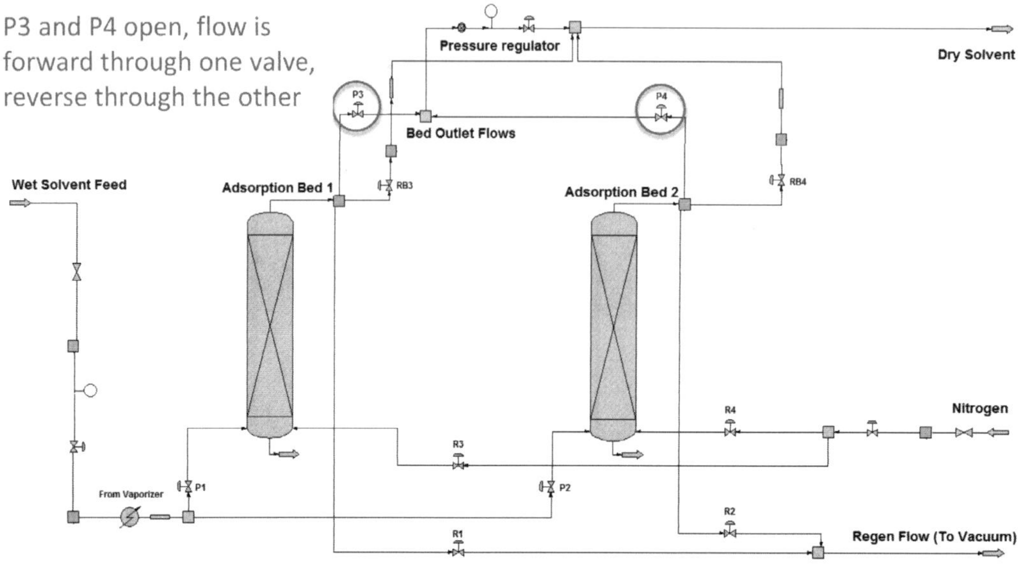

Chemstations™

20

WINTEK Sequence of outlet and regeneration flows out of adsorber beds

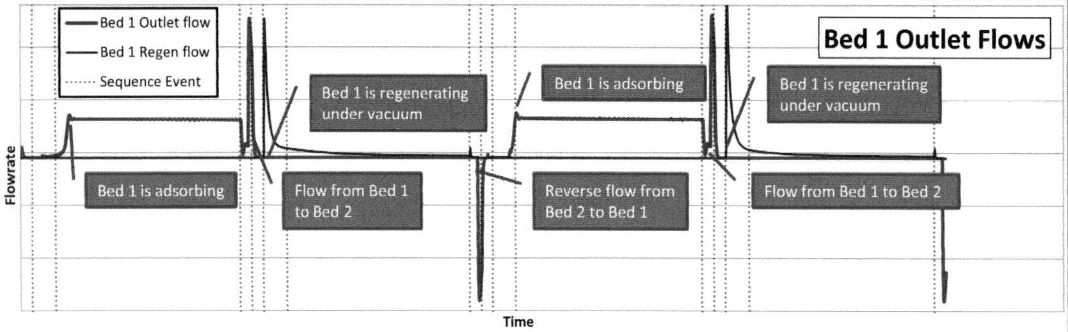

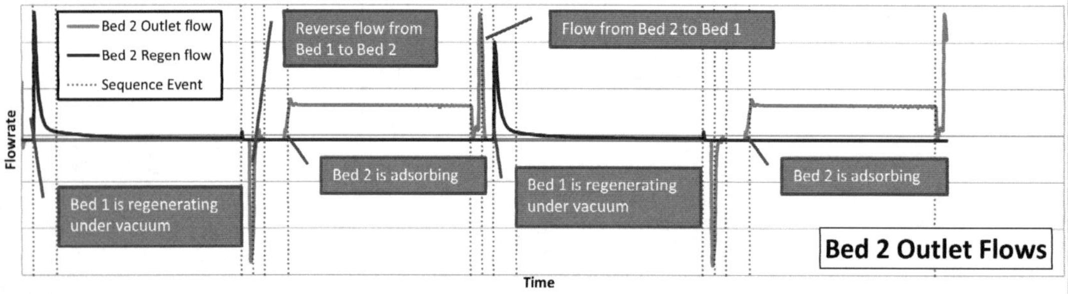

Chemstations™

21

Optimization of schedule (future)

- By using an optimization engine (either the built in engine in CHEMCAD, or an external one connected to the Excel spreadsheet) the schedule of valve events can be optimized to maximize on-stream performance (blue area of chart).

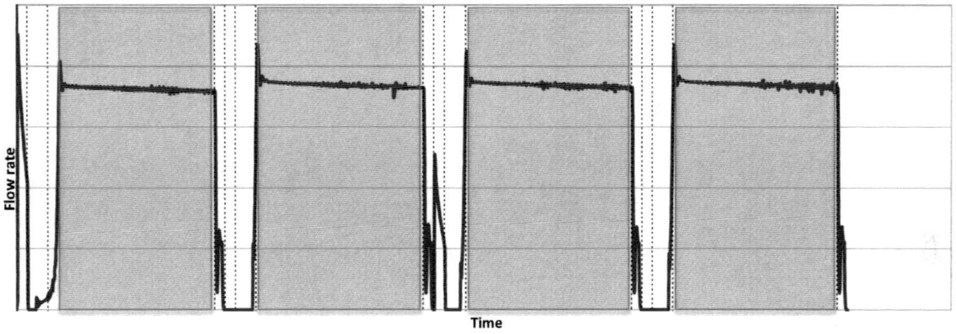

Chemstations™ 22

WINTEK Conclusions

What did we learn from our model?

- Discrete event scheduler with pressure and flow calculations (including reversible flow) allowed optimization of the process schedule
- Ability to identify and quantify
  - Bottlenecks
  - Sequence timing issues
  - Equipment limitations
  - Malfunction effects (bed 1 blowdown valve sticking)
  - Control loop tuning issues
- Ability to use the model for process scaling: process can be scaled up or down to meet customer requirements using the model before building/assembly
- Ability to optimize schedule of events

Chemstations™ 23

Suggested Procedure

1. Start with a simplified model (like CC-BATCH) to get heat duties, initial equipment sizes/specifications, and an initial timing/schedule
2. Build a rigorous dynamic model with as much detail as required (but no more than required) to solve the engineering problem (e.g., only add reversible flow if it is a concern)
3. Build an event sequence control scheme using DATAMAP and Excel starting with the information gathered in Step 1 above. Progressively improve the sequence by running the dynamic model with the scheduler

Chemstations™ 24

Special Thanks

Paul Winter, President and Owner

Wintek Corporation

http://www.wintek-corp.com

230 U.S. Highway 206 STE 401
Flanders, New Jersey 07836

Wintek has been providing equipment solutions to process applications operating at vacuum conditions since 1986, and serves the biodiesel, chemical, environmental, food, pharmaceutical, and plastics industries. Our focus is on process and equipment reliability for your application at a reasonable cost.

Chemstations™ 25

INLINE SYSTEM TO DETECT HYDRATES, WAX AND SAND

IMAGING Thomas M. Canty, P.E.
President, J.M. Canty Inc
Buffalo, NY 14094

ABSTRACT

Inline imaging systems are capable of viewing and digitally analyzing hydrates, waxes and sand in crude oil flow lines at pressure through 10,000 psi and temperatures up through 300 deg C. The technology relies on lenses that are made of fused glass which enable a quality, high speed image that detects and quantifies the size, shape, concentration and color of the suspended components in a crude oil stream. This allows for research and testing in addition to inline measurement and control to assure proper flow

INTRODUCTION

Vision systems are developed to the point where they can provide analysis and control in the oil and gas industry on a reliable and accurate basis. Design of hardware and software, and advancements in Ethernet communication technology have allowed vision based systems to see into process environments such as crude oil to detect and analyze particle constituents. Applications involving the monitoring and control of Produced Water concentrations, Water in Oil analysis, Multi-cut Sand, Oil and Water, Froth (Flotation Control) and Interface can also be effectively handled by vision with the benefit of visual optimization by the operator.

Vision analysis has been proven for many years in applications such as polystyrene bead manufacturing, sugar crystal growth, cell growth, abrasive manufacturing and many others, providing an extra dimension to process analytics that enables the process engineer to make evaluations based on his own expertise and then to use the information to better set software parameters that enable the system to more accurately analyze and control the process.

VISION TECHNOLOGY - HARDWARE

The conditions of the application require a vision system to have the ability to survive in an extreme environment, maintain explosion proof electrical ratings and be easily used by operators in order to be effective. The first point of discussion in this regard deals with the hardware mounted to the process.

Fused Glass Technology

The viewing mechanism of the vision system is most important and must be fitted with rugged, high pressure windows through which illumination and imagery can be conducted. The best option is the fused, glass to metal seal because it is rugged and can withstand pressure, temperature, shock and vibration. In addition, because it is fused directly to the metal probe it can be polished flush with the metal leaving no sites for particle adherence and contamination build up (see Figure 1). The fused glass interface (Fig. 2) is ground and polished to a flat, crevice free surface which remains clean during operation.

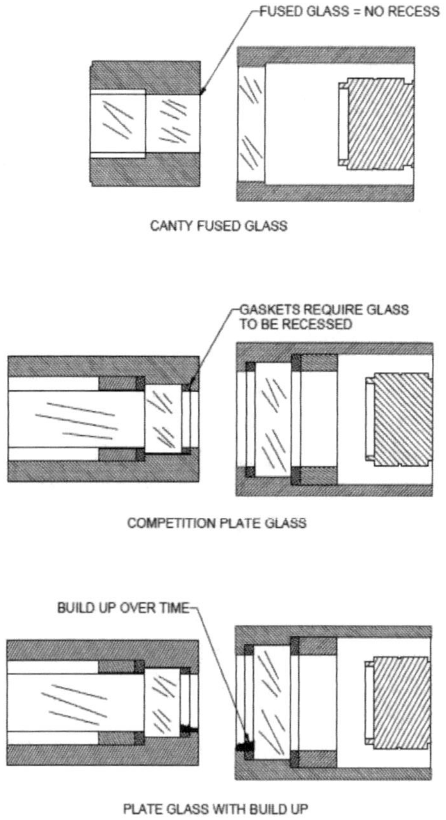

Fig. 1 – Fused Glass vs. Plate Glass Design and Cleanliness

Illumination

Illumination is critical to the success of vision based technology, and a consistent, high intensity source is the building block that allows software to define particles consistently and accurately. Another important feature of a complete system is the ability to control the measurement zone where the particles are illuminated and seen by the camera. The light pipe component shown in Figure 2 has adjustment capability in order that the gap between the camera and light pipe (measurement zone) can be adjusted to optimize the sensitivity of the system. Fluids can range from relatively clear (low turbidity) to quite heavy in particulate matter (high turbidity) and the control of the measurement zone enables the illumination to be controlled by means other than adjustment of intensity from the bulb. A typical light system is shown in Figure 3.

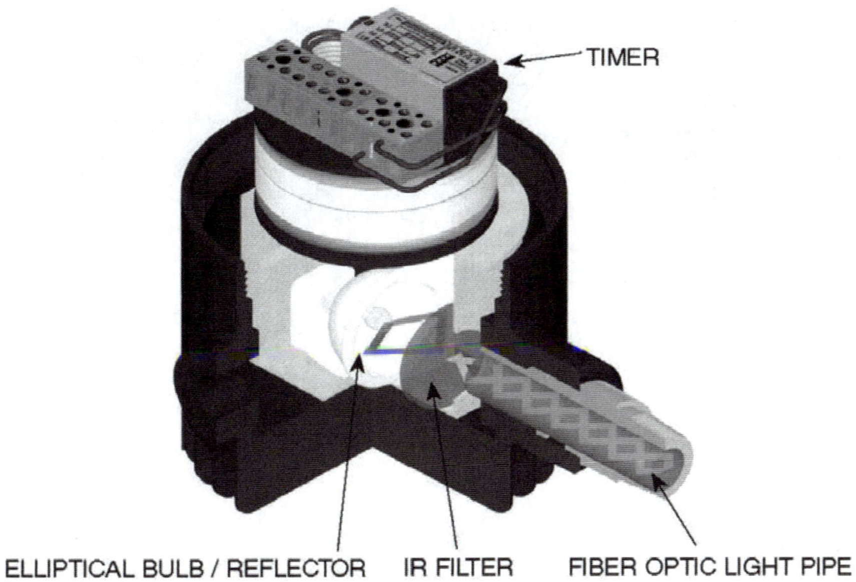

Figure 3. High Intensity Light Section

Cleanliness / Jet Spray Ring Technology

Development of a Jet Spray Ring generates a high pressure vortex rinsing action to remove tough deposits from Vision Systems. Originally designed for the U.S. Space Shuttle program, the Jet Spray Ring accepts a rinsing/cleaning solution and distributes the pressurized material through the specially designed rinse ports. These rinse ports have been engineered to provide maximum cleaning power in a compact package. The Jet cleaning action expels all debris from the Vision System glass, providing an exceptional view into the process. For petroleum applications the spray ports are designed into the instrument probe and maintain the configuration on the measurement zone.

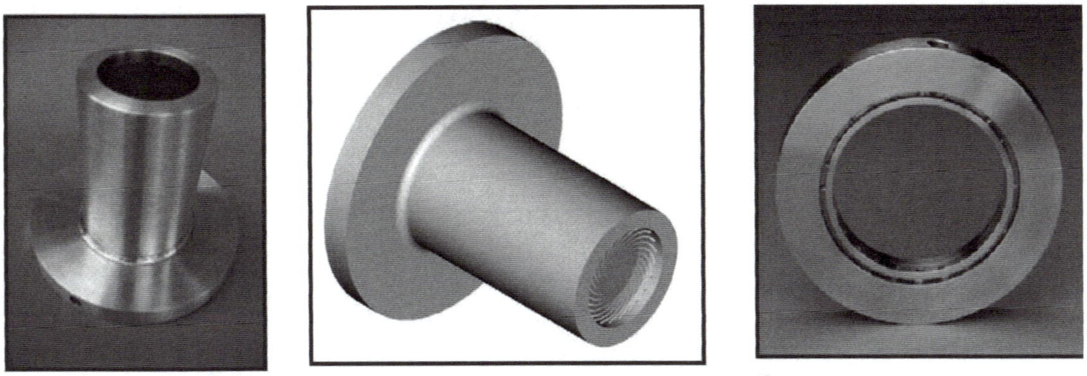

Fig. 4 – Typical Jet Spray Rings

Fig. 5 – Jet Spray Ring Cleaning Action Sequence

Typical Instrument Configurations

Fig. 6 - Inline System for 6" Flow Line and Larger

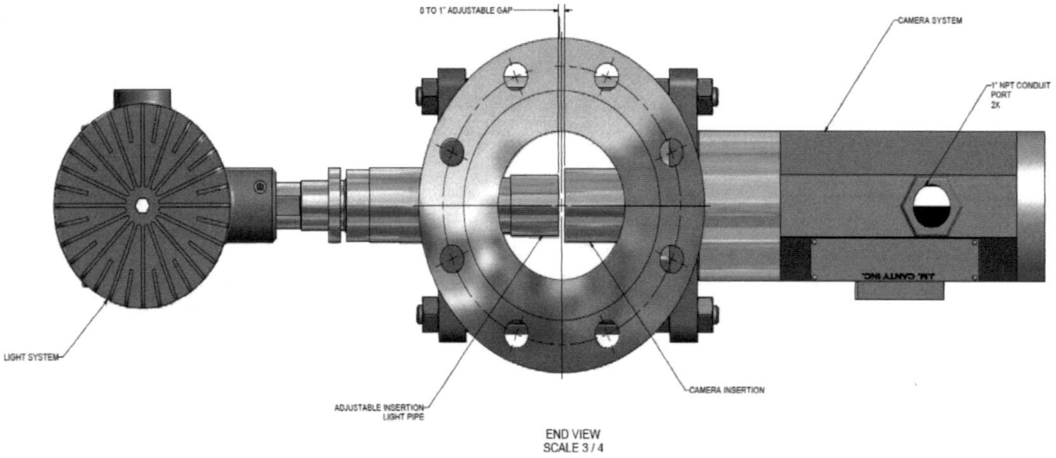

Fig. 7 - Inflow Systems ½" – 4" Pipeline

VISION TECHNOLOGY - SOFTWARE

The second point of discussion deals with the software. Vision systems are generally based on the same concept as the human eye – brain interaction. Objects are seen in our field of vision because they stand out from the background. This delineation is created by differences in light intensity around the object caused by shadows or colors. The vision system detects particles in much the same way. Intensity differences from the background field occur at particle boundaries. The vision system detects these and is able to form the particle based on defined threshold information which sets the level of intensity change that defines a particle boundary. The threshold information is input by the operator based on what he sees on screen. This is the powerful advantage of vision over other types of instrumentation. Reliance on indirect calibration is not required. The actual particles are used to define what is a proper size detection and what is not. This is depicted in the Fig. 8-9.

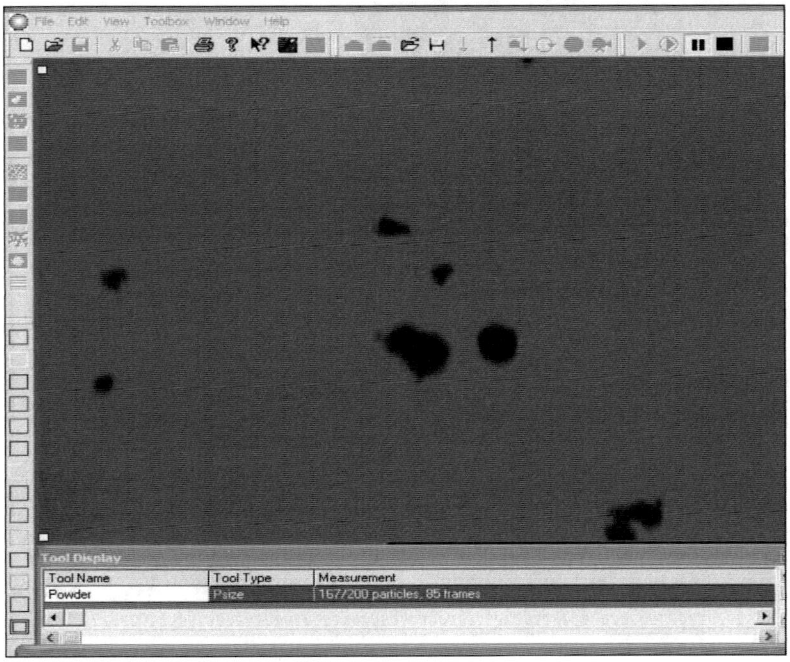

Figure 8. Live Stream View

Figure 8 shows an image of oil particles in the stream which can be used to define to the software what the particles actually are. The next step is to digitize the image and test thresholds to properly define what a particle is based on the actual view of the particle. Figure 9 shows a digitized image of the particle capture which yields particles as they are actually seen. This ability to select the threshold is a way the operator can confirm that the software is capturing particles correctly for analysis.

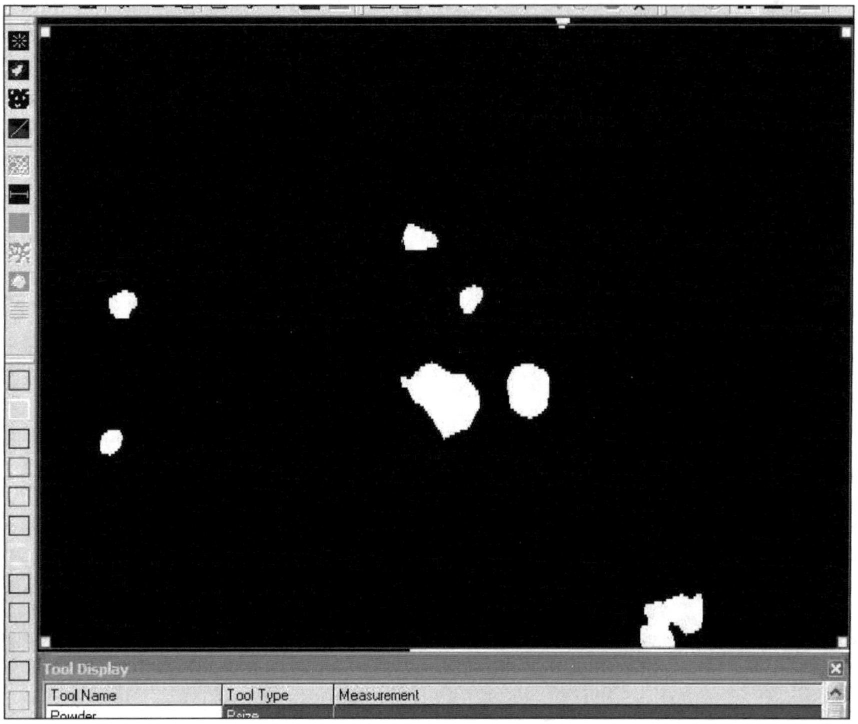

Fig. 9 – Proper Threshold Selected

The discussion to this point will serve as a basis for the remainder of this paper. The hardware and software principles are largely the same regardless of applications. The discussion to follow will focus on the data that can be extracted from the process image to enhance the production and control of the aforementioned applications.

DETECTION OF WAX and HYDRATES

The presence of precipitated wax in a flow line or pipeline can lead to clogging issues. Detection of wax precipitation can alert the operator to the pending problem and prevent a stoppage in production at best, and a loss of a flow line at worst. As process temperature cools, the wax takes on a solid form which tends to adhere to metal surfaces. Build up over time will choke the line. Figure 10 shows results of the system's ability to track wax precipitation in field trials as a function of temperature. Figure 11 shows a typical image of a crude oil which contains water and precipitated wax. The circular ring shapes are water droplets and the semi transparent, large formations are wax particles.

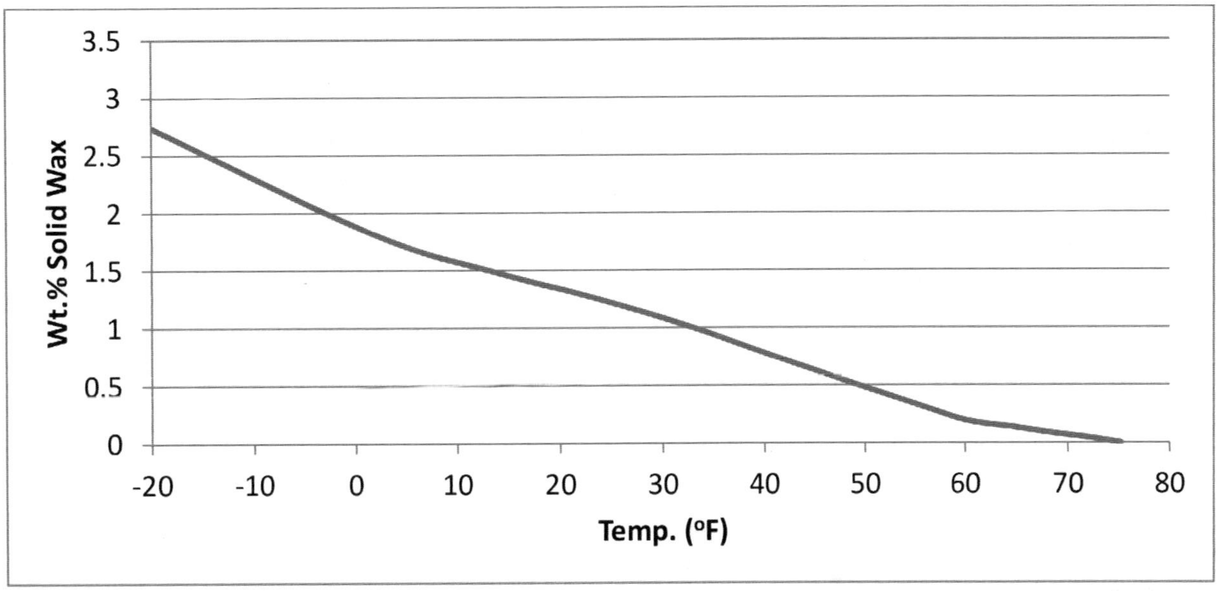

Figure 10. Wax vs Temperature

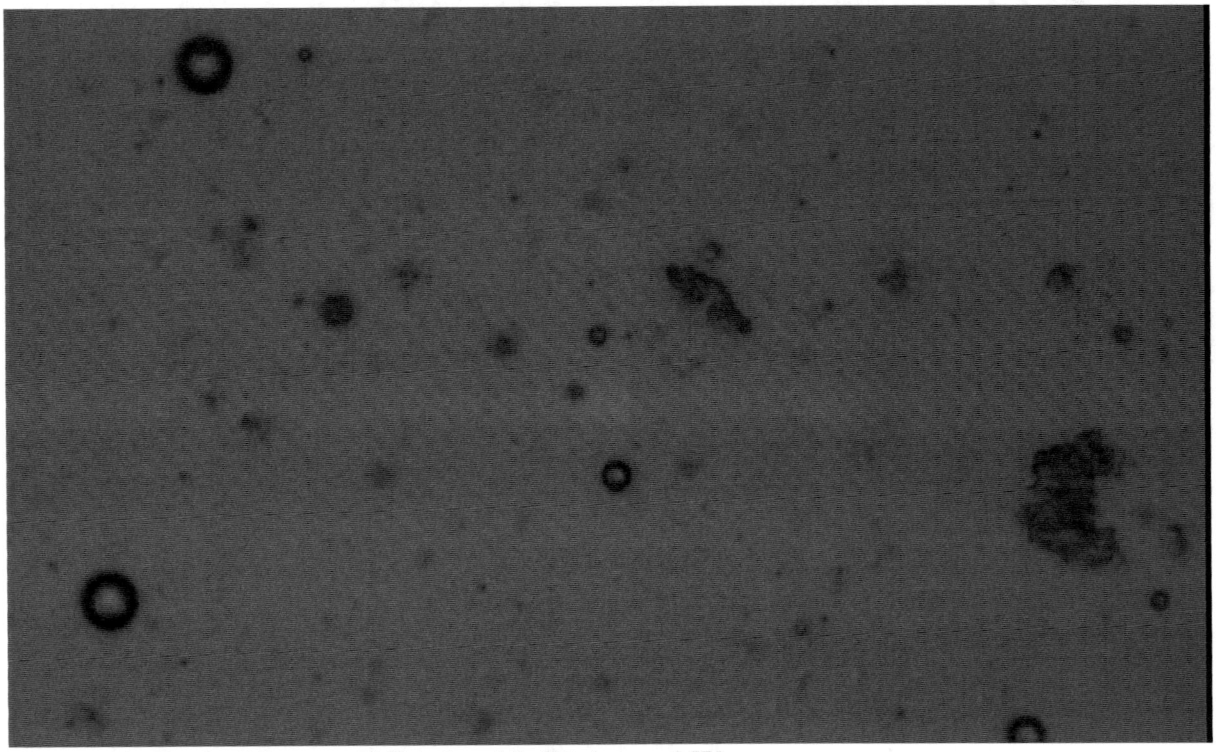

Figure 11. Typical Image of Crude with Water and Wax

Hydrates are detected in a similar fashion as they crystallize out of solution. Methane droplets will start to freeze, then agglomerate and finally plug the line if left untreated long enough by either temperature control or chemical injection. As their formation proceeds they will obtain an opaque shell which is indicative that the process is well underway and action must be taken to avoid disaster.

DETECTION OF SAND

The detection of sand is a critical function in the separation process. Once removed from the oil it must be cleaned, injected or dumped if present in small enough quantities. Detection is critical in any case. Viewing a produced water image one will see oil droplets, sand and other contaminants. General acceptance levels to dump to the environment are 30 ppm which is not a lot. When sand is detected in the outflow of the separator it is an indicator that separation must be adjusted to handle the present characteristics of the crude. Unlike oil droplets, sand, and other solids will almost always be non circular in the view. Vision software allows the categorization of these shapes for identification. In the image following a sand particle is seen in the center and upper left corner. These can be identified by aspect ratio or circularity as an example. Concentration of each constituent can be determined.

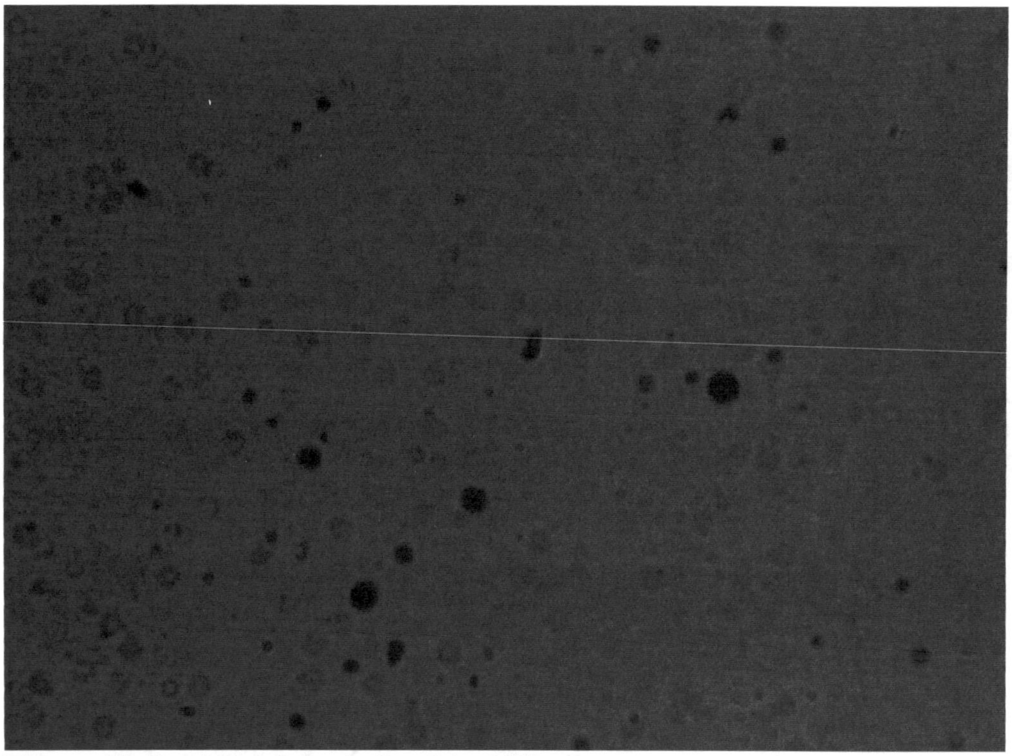

Figure 12. Sand Particles in Produced Water

CONCLUSION

Vision technology has the ability to maintain its viability in the harsh oil environment through design and manufacturing inherent in its components. Good visual optics, lighting and control of the measurement zone are all critical elements to success. Software has the advantage over other technologies due to its capability to analyze shape as well as size. This makes the difference between counting particles and classifying particles for a full understanding of how the process is behaving.

*** Formatting Issue - Best Available Paper/Graphic ***

Solid State Carbon Sequestration by Utilization of Agro-wastes to Produce Value-added Products

Soh Fong Lim[1], Mohammad Omar Abdullah[1], Andrew Ragai Henry Rigit[2]

[1] Department of Chemical Engineering and Energy Sustainability, University Malaysia Sarawak, 94300 Kota Samarahan, Malaysia

[2] Department of Mechanical and Manufacturing System Engineering, University Malaysia Sarawak, 94300 Kota Samarahan, Malaysia

Abstract

Solid state carbon sequestration via utilization of agro-wastes to produce value added products such as biofertilizer and biosorbent is studied in this paper. The carbon sequestration is a sustainable development process which can be used to curb the negative impacts of human activities by reducing the emission of CO_2 via locking the carbon in solid form. Analytical analysis has found that both of the products contain relatively high percentage of organic carbon compare to other organic material such as soil. These might be due to the locking of carbon in solid form during the application of these products.

Keywords: agro-wastes, biofertilizers, biosorbents, solid state carbon sequestration.

I. INTRODUCTION

Around the globe, waste is piling up at an alarming rate and we need an effective waste disposal schemes very urgently while at the same time making use of the unwanted waste to provide clean energy or new materials via smart and efficient composting or recycling methods. The second important issue is linked to the goal towards global sustainability in terms of reducing greenhouse gas emissions, in particular CO_2 (Abdullah, 2012).

Solid state carbon sequestration is used to reduce carbon footprint and to minimize waste. The carbon footprint is formed by the existence of carbon dioxide (CO_2) gas in the atmosphere. CO_2 which is well known for driving global warming and climate change is the most important greenhouse gas emitted by industries and human activities. In this instance, agricultural activities are increasingly being required to account for its contribution to the release of CO_2. Thus, sustainability in the agricultural activities is important to conserve the natural environment and resources for our future generation. The sustainability should also extend beyond the economic well-being of the current generation and reflects the ability of future generations to meet their needs. These could be achieved by steering the agricultural activities along a more sustainable path using green technology or sustainable technologies.

Carbon sequestration is a sustainable development process that removes carbon footprint. This process can be used to curb the negative impacts of human activities by reducing the emission of CO_2 via locking the carbon in solid form such as in trees, soils, and other value-added solid products (eg. bio-fertilizer, bio-sorbent, etc.). There are numerous agro-wastes being produced in plantation and processing such as skins/peels, cores, pits, leaves, brunches, and pericarp. Agro-wastes are recognized as having hidden economic values which could be benefited by transforming the agro-wastes into useful products such as biofertilizer and biosorbent (Fortuna et al., 2003; El-Haggar, 2007; Kukal et al., 2009). Rivera-Cruz et al. (2008) reported that banana wastes are effective biofertilizer carriers for promoting plant growth and soil sustainability. Sud et al. (2008) reviewed the potential of agricultural waste materials in producing biosorbent for sequestering heavy metal ions from aqueous solutions. These agro-wastes largely compose of hemicelluloses, lignin, extractives, lipids, proteins, simple sugars, water hydrocarbon, starch, and etc. (Sud et al., 2008). Thus, these agro-wastes will be targeted for potential green technology application and sustainable development/practices.

2. MATERIALS AND METHODOLOGY

2.1 Materials

Banana peels (*Musa Acuminata*) that were used in this study were agro-wastes that were produced from local industry.

2.2 Biofertilizer

Composting method was used to produce biofertilizer from banana peels. 500 kg of banana peels were placed in a blender with 200 ml of water. Then, the blended banana peels were exposed to the sunray for 2 days (dehydration process). The dried banana peels were then put into a bin together with 50 g of sawdust, 250 g of lignite, and 25 ml of nipah palm sugar. The mixture was stirred for 5 minutes. After that, the bin was lid tightly and kept for 3 weeks for composting process purposes. Within the 3 weeks of composting process, compost pile was stirred frequently to allow the mixture to be well mixed and reach optimum temperatures as well as not overheat. After 3 weeks of the composting process, all the compost materials were removed from compost bin and the dehydration process was repeated for 4 hours. This is to ensure that the biofertilizer are in dry solid state (Matu, 2010; Muhammad, 2011).

2.3 Biosorbent

The banana peels were washed thoroughly with tap water to remove the adhering dirt and dried in a laboratory oven at 60°C for 58 hours. Then, the dried banana peels were crushed into a powder form using a laboratory grinder. The powdered banana peels were sieved to obtain a uniform size of the crushed banana peel. Finally, the powdered banana peels were thoroughly washed with deionized water to remove physically adsorbed contamination and dried in a laboratory oven at 80°C for a period of 24 hours (Sugumaran, 2010; Meikle, 2011).

3. RESULTS AND DISCUSSION

3.1 Value-added Products

Agro-wastes from banana industry mainly banana skins are used to produce two types of useful products; biosorbent and biofertilizer (Figure 1). These products are fabricated using simple and cost-effective methods. Both of the products have shown significant results in their applications. The biosorbent is able to remove crystal violet dye from aqueous solution. The maximum uptake of the crystal voilet dye is 245.74 mg/g. Removal of the dye is very fast, as it reached equilibrium in 3 hours of contact time. It is also found that the removal of the dye is highly pH dependent. Within the tested pH range (pH 2 to 12), the maximum removal of the dye is between pH 6 and pH 8. On the other hand, the biofertilizer which was fabricated using solid state fermentation method is able to increase the yield of crop which was proven by the plantation of the mustard and water spinach vegetables.

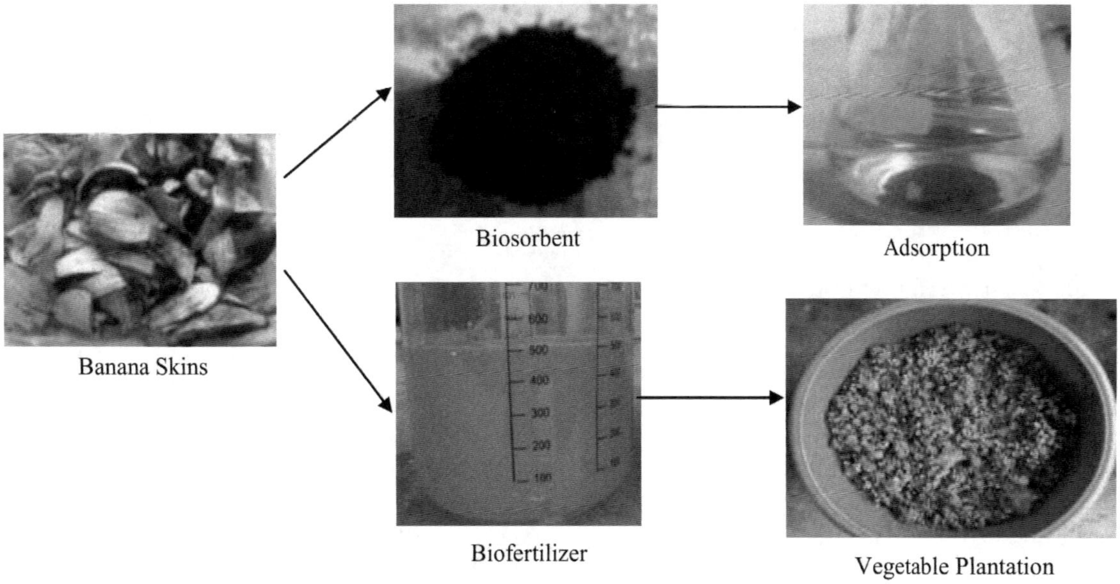

Figure 1 Flow diagram for value-added products from banana skins

3.2 Carbon analysis

Samples from the biosorbent and biofertilizer were analyzed to determine the amount of carbon content using Walkley-Black method (Saarani, 2011). Walkley-Black method is chosen based on its simplicity and rapidity (Islam and Weil, 1998; Schumacher, 2002; Krishan *et al.*, 2009). Walkley-Black method is wet combustion method where the amount of carbon is determined by the reaction of $Cr_2O_7^{2-}$ and H_2SO_4 without external heating. In this method, the amount of organic carbon content in the sample is determined from the difference of amount $FeSO_4$ used in the blank titration with the amount of $FeSO_4$ used for reaction with the remaining un-reacted part of $K_2Cr_2O_7$. The chemical reaction that involved in Walkley-Black methods is shown in equation (1).

$$2\,Cr_2O_7^{2-} + 16H^+ + 3\,C \rightarrow 4\,Cr^{3+} + 8\,H_2O + 3\,CO_2 \qquad (1)$$

Desired amount of sample (W g) was placed in the 250mL Erlenmeyer flask. 5.0mL of 1.0*N* $K_2Cr_2O_7$ solution is added into the flask that containing the sample and the flask is swirled to mix the sample with $K_2Cr_2O_7$ solution. Then, 10.0mL of concentrated H_2SO_4 is added rapidly to the mixture and followed by adding 100mL of deionized water. The suspension is filtered out and ferroin indicator solution is added into the flask containing the solution. Then the solution is titrated with 0.5N $FeSO_4$ from 50 mL burette. As the endpoint approach, the solution takes on green cast and change into dark blue green. Then additional 0.2 mL of indicator is used to sharpen the endpoint. At this point, the $FeSO_4$ solution is added drop by drop until the colour change from blue green into brownish (maroon in reflected light) and the amount of ferrous sulfate solution used is recorded (*T* mL). The experiment is run again with blank and the amount of FeSO4 solution used is recorded as (*B* mL). The amount of

***** Formatting Issue - Best Available Paper/Graphic *****

$FeSO_4$ solution in the back titration which is obtained from the analytical procedure is then be used to calculate the amount of carbon in the sample in percentage using equation (2).

Where,

- B = the amount of $FeSO_4$ solution used to titrate blank
- T = the amount of $FeSO_4$ solution used to titrate sample
- N = the normality of $FeSO_4$ solution
- W = weight of the sample in mg
- 3 = equivalent weight of carbon in mg.
- y = oxidation factor

Carbon analysis is performed on both of the value-added products. Table 1 demonstrates the percentage of carbon in biofertilizer using Walkley-Black analytical method. Based on Table 1, the amount of carbon in the biofertilizer is approximately 30.09 – 35.10 wt%.

Table 1 Amount of $FeSO_4$ used in back titration and percentage of carbon in biofertilizers

SAMPLE	Total amount of $FeSO_4$ used in back titration (mL)	Percentage of carbon (wt%)	
		y = 1.14	y = 1.33
Biofertilizer 1	5.60	31.16	36.35
Biofertilizer 2	2.58	29.01	33.85
Average		30.09	35.10

Figure 2 shows an increase of carbon content in biosorbent at different crystal voilet (CV) dye concentration. For both oxidation factors, the amount of carbon in the biosorbent was significantly increased as the dye concentration increased until it reached saturated state which is at CV dye concentration of 120 mg/L and above. This saturated state shows that the biosorbent is not able to absorb the CV dye anymore. Thus, the adsorbed carbon which is in CV dye structure remains constant when the adsorptions reach saturated state (Andersen *et al.*, 2009). Manocha *et al.* (2011) also stated that the formation of sample sizes and the interconnectivity between adsorbate and adsorbents can affect the rate of adsorption capacity. The weight of the biosorbents also can affect the percentage of absorption of the dye where as the weight of biosorbents increased, the absorption rate also increased (Raymundo, 2010).

*** Formatting Issue - Best Available Paper/Graphic ***

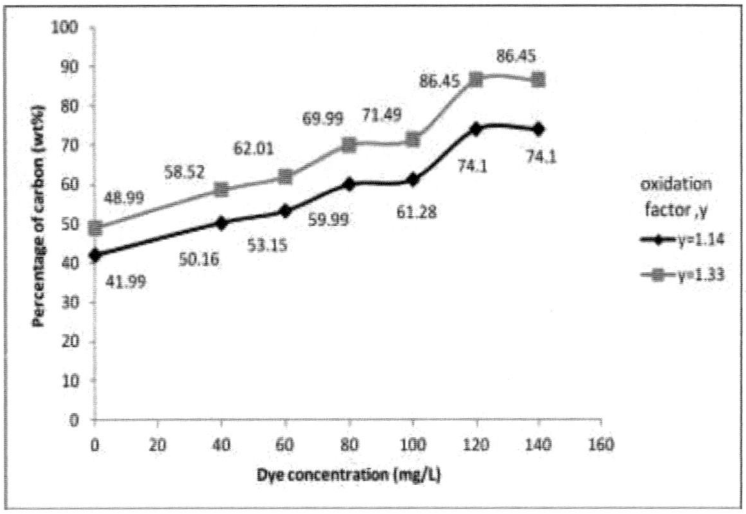

Figure 2 Percentage of carbon versus dye concentration in (mg/L)

Thus, analytical analysis has found that both of the useful products contain relatively high percentage of organic carbon compare to other organic material such as soil (Kim, 2005). Thus, the carbon is locked in solid form which could reduce the emission of CO2. This process is called carbon sequestration which is a sustainable development process to curb the negative impacts of human activities.

ACKNOWLEDGEMENT

The authors would like to thank Universiti Malaysia Sarawak for the use of laboratory facilities and equipments. This project is funded by Fundamental Research Grant Scheme (FRGS/02(17)/738/2010(24)) from the Ministry of Higher Education, Malaysia.

REFERENCES

A. Fortuna, R. Harwood, K. Kizilkaya, E. A. Paul, "Optimizing nutrient availability and potential carbon sequestration in an agroecosystem." *Soil Biology and Biochemistry*, Vol. 35, 1005-1013, 2003.

A. S. Raymundo, R. Zanarotto, M. Belisário, and M. de Godoi, "Evaluation of Sugar-Cane Bagasse as Bioadsorbent in the Textile Wastewater Treatment Contaminated with Carcinogenic Congo Red Dye," *Brazilian Archives of Biology and Technology* ,vol. 53, no.4, pp. 931-938, 2010.

B. A. Schumacher, *Methods For Determination of Total Organic Carbon (TOC) in Soils and Sediments* , pp. 1-18, 2002.

D. Sud, G. Mahajan, and M. P. Kaur, "Agricultural Waste Material as potential adsorbent for sequestering heavy metal ions from aquenous solution – A review." *Bioresource Technology*, vol. 99, no. 14, pp. 6017-6027, 2008.

F. S. Saarani, "Elucidation of Carbon Content via Laboratory Analysis," Bachelor of Mechanical and Manufacturing Engineering Thesis, Department of Mechanical and Manufacturing Engineering, Faculty of Engineering, Universiti Malaysia Sarawak, 2011.

G. Krishan, S. K. Srivastav, S. Kumar, S. K. Saha, and V. K. Dadhwa, "Quantifying the underestimation of soil organic carbon by the Walkley and Black technique – examples from Himalayan and Central Indian soils," *Current Science* , vol. 96, 2009.

H.T. Kim, *Soil Sampling, Preparation, and Analysis* (2nd Edition ed.). Boca Raton: Taylor & Francis Group, 2005.

K. R. Islam and R. R. Weil, *A Rapid Microwave Digestion Method for Colometric Measurement of Soil Organic Carbon*, 1998, pp. 2269-2284.

M. C. Rivera-Cruz, A. T. Narcía, G. C. Ballona, J. Kohler, F. Caravaca and A. Roldán, "Poultry manure and banana waste are effective biofertilizer carriers for promoting plant growth and soil sustainability in banana crops." *Soil Biology and Biochemistry*, Vol. 40, pp. 3092-3095, 2008.

M.O. Abdullah, *Applied Energy: an introduction.* New York: CRC Press (Taylor & Francis Group), 2012.

N. Muhammad, "Evaluation of Bio-Fertilizer from Agricultural Waste," Bachelor of Mechanical and Manufacturing Engineering Thesis, Department of Mechanical and Manufacturing Engineering, Faculty of Engineering, Universiti Malaysia Sarawak, 2011.

S. Manocha, J. Bhagat, M. Patel, N. Patel, and L.M. Manocha, Adsorption Behaviour of Carbons from Bio-Mass, pp. 727-731, 2011.

S. M. El-Haggar, "Sustainable Industrial Design and Waste Management." Academic Press, Chapter 7, pp223-260, 2007.

S. U. Matu, "Utilization of Agro-waste to Produce Biofertilizer," Bachelor of Mechanical and Manufacturing Engineering Thesis, Department of Mechanical and Manufacturing Engineering, Faculty of Engineering, Universiti Malaysia Sarawak, 2010.

S. S. Kukal, Rehana-Rasool, D. K. Benbi, "Soil organic carbon sequestration in relation to organic and inorganic fertilization in rice–wheat and maize–wheat systems." *Soil and Tillage Research*, Vol. 102, pp. 87-92, 2009.

T. A. Sugumaran, "Utilization of Agro Waste to Produce Biofertilizer," Bachelor of Mechanical and Manufacturing Engineering Thesis, Department of Mechanical and Manufacturing Engineering, Faculty of Engineering, Universiti Malaysia Sarawak, 2010.

V. Meikle, "Removal of Contaminants Using Biosorbent," Bachelor of Mechanical and Manufacturing Engineering Thesis, Department of Mechanical and Manufacturing Engineering, Faculty of Engineering, Universiti Malaysia Sarawak, 2011.

W. C. Andersen, S. B. Turnipseed, C. M. Karbiwnyk, R. H. Lee, S. B. Clark, W. D. Rowe, M. R. Madson, K. E. Miller, Quantitative and Confirmatory Analyses of Crystal Violet (Gentian Violet) and Brilliant Green in Fish. *U.S. Food and Drug Administration, Animal Drugs Research Center,* 2009.

(105b) Ni-CaO-ZrO2 Nanocomposite Catalyst for CO2 Reforming of CH4

Qingjun Chen[1], Jun Zhang[1], **Tiejun Zhao**[1] and Yuhan Sun[2], (1)Low Carbon Conversion Center, Shanghai Advanced Research Institute, CAS, Shanghai, China, (2)State Key Laboratory of Coal Conversion, Institute of Coal Chemistry, Chinese Academy of Sciences, Taiyuan, China

1. 1. Introduction

Recently, utilization of natural gas to fuels and chemicals receives more attention due to its abundance over the world and less CO_2 emission. The production of syngas by CO_2 reforming of methane is one of the attractive routes to reach this goal. Compared with conventional steam reforming, CO_2 reforming reaction tends to easier coking formation and needs extra 20% heat to obtain similar conversion. Nickel-based catalysts have shown an excellent behavior in this reaction with a comparable activity to noble metal catalysts [1]. However, coke formation and sintering of nickel particles at a high temperature condition are often observed on Ni catalysts [2]. Developing a high-temperature stable catalyst is highly interesting for this reaction by controlled synthesis of catalyst. In this contribution, a stable Ni-CaO-ZrO_2 nano-composite catalyst for CO_2 reforming of CH_4 is prepared by a coprecipitation method. The effect of preparation condition is carefully investigated. The relationships between catalyst structures and activity of nanocomposite catalysts are discussed.

1. 2. Experimental

Ni-CaO-ZrO_2 catalysts were prepared by a coprecipitation method with a reflux-digestion process. After precipitation, the obtained slurry was washed and refluxed in de-ionized water, then filtered, dried and calcined at 600 ^oC for 5 h. The catalysts prepared at different reflux-digestion condition were assigned as NC-X-Y, where X and Y are the reflux temperature and time, respectively. The catalyst without reflux was assigned as NC-0. The reforming reaction was conducted at 850-900 °C in a quartz fixed-bed reactor at atmospheric pressure. Obtained catalysts are tested by feeding the reaction mixture of CO_2 and CH_4 (molar ratio 1.2: 1) without dilution and a gas hourly space velocity (GHSV) of 79 000 ml/(hg) . In addition, comparison of calciantion procedure and coking quantity was performed at different operating conditions.

1. 3. Results and discussion

3.1 Physical and chemical properties of the Ni-CaO-ZrO_2 catalysts

The pore structure parameters are collected in Table 1. The catalyst without reflux-digestration has the smallest specific surface area (63 m^2/g) and pore volume (0.09 cm^3/g). With the increase of the reflux time, the specific surface area and pore volume of the catalysts increased. NC-100°C-24h exhibits the largest surface area of 245 m^2/g. Reflux temperatures also have a significant effect on the pore structures of the catalysts. At the same reflux time, the specific surface areas of the catalysts increased with increasing the reflux temperature.

Table 1. Pore structure parameters of the catalysts

Samples	S_{BET} m^2/g	V_t cm^3/g	D_m nm

*** Formatting Issue - Best Available Paper/Graphic ***

NC-0	63	0.09	5.58
NC-100°C-6h	152	0.18	4.73
NC-100°C-12h	176	0.21	4.85
NC-100°C-24h	245	0.30	4.95
NC-60 °C-24h	92	0.13	5.83
NC-80 °C-24h	160	0.21	5.31

The XRD patterns of the catalysts prepared at different reflux-digestion condition are shown in Figure 1. The catalyst without reflux has strong diffusion peaks of ZrO_2 and NiO. As the increase of the reflux time and temperature, diffusion peaks assigned to ZrO_2 and NiO became weaker and wider. This result indicates that reflux digestion improves the dispersion of NiO and decreases the size of ZrO_2 crystals in the catalysts.

TPR profiles of Ni-CaO-ZrO_2 catalysts are illustrated in Figure 2. The peak or shoulder at about 400 °C is attributed to the reduction of large NiO particles, while the peaks at 600 °C and 700 °C may belong to the reduction of NiO that have the weak and strong interactions with ZrO_2, respectively. For the catalyst without reflux, NiO mostly has the weak interaction with the carrier or is in the form of bulk particles. At the reflux temperature of 100 °C, the interaction between NiO and ZrO_2 became stronger with the increase of reflux times (Figure 2 a). However, when the reflux temperature is lower than 100 °C, the interaction between NiO and carrier was not significantly enhanced (Figure 2 b). So the high reflux temperature and long reflux time are key factors to prepare the catalysts with strong interaction between metal and the carrier.

Figure 1. XRD patterns of the catalysts Figure 2. TPR profiles of the catalysts

3.2 **Activity of the Ni-CaO-ZrO_2 catalysts for CO_2 reforming of methane**

Results of the catalytic tests at 850 °C with a high GHSV (CO_2/CH_4=1.2:1) of 79, 000 ml/(h·g) are shown in Figure 3(a). The catalyst without reflux has a low activity with the CH_4 conversion of about 70 %. In addition, CH_4 conversion of this catalyst decreased with time and only 55% (conversion) is achieved after reaction on stream for 150 h. A higher conversion of CH_4 and stability (very close to equilibrium) is observed on the catalyst refluxing at 100°C for 24 h. Therefore, prolonging the reflux time will increase activity of Ni-CaO-ZrO_2 catalysts for CO_2 reforming of methane. The catalyst with 24 h reflux-digestion (NC-100 °C-24h) exhibits a highly stable activity over 1000 h without obvious deactivation at a GHSV of 79, 000 ml/(hg) at850 °C (Data are not shown) .

NC100-24 h

*** Formatting Issue - Best Available Paper/Graphic ***

Figure 3. Catalytic performance of catalysts for CO_2 reforming of methane

(a) Effect of reflux temperature and time; (b) Effect of calcination temperature; (c) Effect of calcination procedure

Furthermore, the calcination temperature effect on the catalyst activity is carried out at similar operating conditions. It is found that the activity decreases when the sample is calcined at static air at a higher temperature. The NC-100-24h sample calcined at 900°C presents a very low activity (Fig.3b) . XRD examination indicates that monoclinic ZrO_2 is gradually formed and particle size of ZrO_2 becomes larger. TPR measurement indicates no obvious peaks for NiO reduction for the sample calcined at 900°C. These observations indicate that the sintering occurs at high temperatures. However, if the catalyst is directly reduced in the mixtures of H_2 and N_2 at 850°C or 900°C for 3 h, the conversion of CH_4 still can maintain a high CH_4 conversion (Fig. 3c), implying that the calcination procedure has a profound impact on the activity of catalyst. Interestingly, the catalyst performing at 900°C gives a lower carbon deposition than 850 and 800°C after time on stream for 200 h. The full operating window will be given by a more systematic study afterwards.

1. 4. Conclusions

A high-temperature stable Ni-CaO-ZrO_2catalyst is developed by the coprecipitation method with a reflux-digestion process. Reflux digestion can improve the dispersion of NiO and enhance the strong interaction between metal and support. The obtained Ni-CaO-ZrO_2 catalyst can be further improved by modifying the calciantion procedure.

References

[1] J. R. Rostrup-Nielsen, J. H. B. Hansen, ***Journal of Catalysis*** 144 (1998).

[2] C. J.Liu, J. Y. He, J. J. Jiang, Y. X.Pan, ***ChemCatChem*** 3 (2011)

(105c) Gas Cleanup Using Elastic Layered Metal-Organic Framework Adsorbents

Francisco Sotomayor and **Christian M. Lastoskie**, Civil & Environmental Engineering, University of Michigan, Ann Arbor, MI

The suitability of elastic layered metal-organic frameworks (ELM) for the removal of carbon dioxide from methane-rich gas mixtures was evaluated using molecular modeling and experimental measurements for two adsorbents, ELM-11 $[Cu(BF_4)_2(bpy)_2]$ and ELM-12 $[Cu(OTf)_2(bpy)_2]$. These materials are promising absorbents for CO_2 separation from biogas and natural gas mixtures, due to their reversible gated isotherm features and their high selectivity adsorption of CO_2 over CH_4 (S = 50-100 at 298 K) from gas mixtures with CH_4 and CO_2 compositions representative of biogas / natural gas extraction streams.

Grand canonical Monte Carlo (GCMC) simulation was used to calculate the framework adsorption capacity and CO_2 selectivity from binary CH_4 / CO_2 mixtures. Simulated CO_2 capacities above the gating pressure and CO_2 isosteric heats of adsorption agree well with ELM-11 and ELM-12 experimental data. Ideal adsorbed solution theory (IAST) estimates of CO_2 selectivity are generally in good agreement with the GCMC simulations, although modifications of the IAST methodology are needed in order to take into account the co-adsorption condition for CH_4 uptake by CO_2 gate-opening. Density functional theory (DFT) calculations are also presented for CO_2 and CH_4 interactions with the ELM-11 and ELM-12 framework structures. Upon the adsorption of CO_2 in the ELM framework, the metal-anion bond of the framework was extended, whereas the same bond length is virtually unchanged upon CH_4 adsorption. DFT analysis of the adsorption binding energies of other trace components in natural gas and biogas production (e.g. H_2S, H_2O, N_2) and CH_4 will also be presented and discussed.

(105d) Effect of Preparation Parameters On Mixed Amine Functionalized Silica Sorbents for CO2 Capture

Adefemi Egbebi[1], Sonia Hammache[1] and McMahan L. Gray[2], (1)URS-NETL, Pittsburgh, PA, (2)U.S. Department of Energy, National Energy Technology Laboratory, Pittsburgh, PA

Effect of preparation parameters on mixed amine functionalized silica sorbents for CO_2 capture

Adefemi Egbebi[1, 2], Sonia Hammache[1, 2], and McMahan L. Gray[1]

[1]US DOE-National Energy Technology Laboratory, Pittsburgh, PA

[2]URS, Pittsburgh, PA

Global climate change has been linked to growing atmospheric concentrations of carbon dioxide. Therefore the development of practical methods to remove, reduce or prevent the release of anthropogenic CO_2 is prudent to slow this increase. The most mature technology for the separation of CO_2 from flue gas streams from power plants is via liquid amine scrubbing which is considered expensive and energy-intensive. The potential therefore exists for the development of alternative CO_2 removal technologies, including the use of solid amine-based sorbents among others, capable of less energy-intensive CO_2separation.

Previous work in our laboratory investigated co-impregnated mixed amine (Polyethyleneimine (PEI) + 3-aminopropyltriethoxysilane (APTES) / Silica) solid sorbents with the goal of improving sorbent performance in terms of its CO_2 uptake capacity, thermal and hydrolytic stability as well as the kinetics during sorption and regeneration cycles. Results showed improved CO_2capture capacity and stable performance of the mixed amine sorbent relative to a single amine (PEI) sorbent.

Current work investigates the effects of synthesis parameters/variables such as bath temperature, solvent type and impregnation sequence on sorbent performance. Modified Pressure Swing Adsorption (PSA) tests were conducted with a thermogravimetric analyzer (TGA) on two sorbents with same composition of 15 wt. % PEI: 15% APTES: 70 % SiO_2 but with different solvents used during their syntheses - methanol for one and methanol-hexane mixture for the other. Adsorption was performed using dry 10% CO_2 /N_2 feed gas and desorption was conducted with 100% N_2 feed gas while temperature was kept the same at 100°C. Preliminary results show negligible differences between the sorbents in terms of CO_2uptake capacity and thermal stability.

The effect of various preparation parameters will be elucidated from the results of TGA, fixed bed reactor and other characterization experiments on sorbents and will be presented.

CO_2 Reduction Efforts for Canadian Oil Sands Industry

Ravishankara Subraya, Oscar Aguilar

The Quest project is a joint venture of the Athabasca Oils Sands Project partners (Shell Canada, Chevron Canada Limited, and Marathon Oil Canada Corporation) to fully integrate CCS technology to capture, transport, inject, store, and monitor CO_2 from the Scotford Upgrader in the Industrial Heartland Region near Fort Saskatchewan, Alberta. CCS is an important part of both Canada's and Alberta's CO_2 reduction plans. Project is currently under execute phase. This facility will sequester 1.08 Mtpa of CO_2 on completion in Q1 2015. The technical presentation will cover technology selection, process & utility integration challenges and key learning from the project.

An Introduction to the America Invents Act

David W. Holt
256-517-5191 dholt@babc.com

BRADLEY ARANT BOULT CUMMINGS LLP

Agenda

- Basics of Patent Law
- The America Invents Act
  - First Inventor to File
  - Other Changes
    - Prior Use Defense
    - Best Mode
    - Post-Grant Proceeding
    - False Marking
    - Joinder

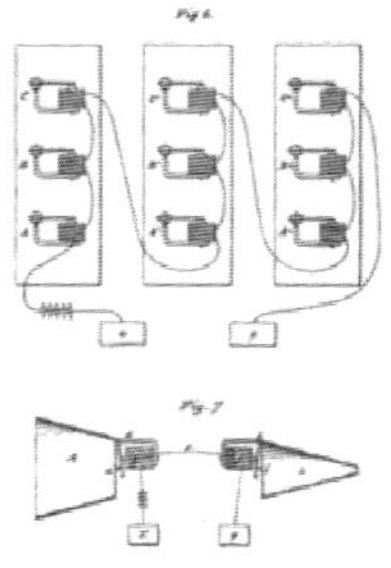

Pat. 174,465 (1876)
Alexander Graham Bell

What is a Patent?

- A patent grants the owner the right to exclude others from making, using, selling, offering for sale, and importing the subject matter bounded by the claims of the patent.
- Period of exclusivity is 20 years from the filing date.*

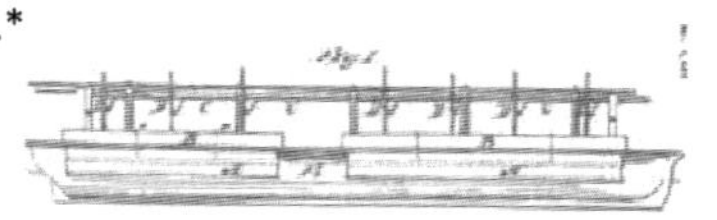

Pat. No. 6,469 (1849): Manner of Buoying Vessels
Abraham Lincoln

Requirements for Obtaining a Patent

- "Process, machine, manufacture, or composition of matter" (Sec. 101)
- Novel (Sec. 102)
- Not obvious (Sec. 103)
- Useful (Sec. 101)

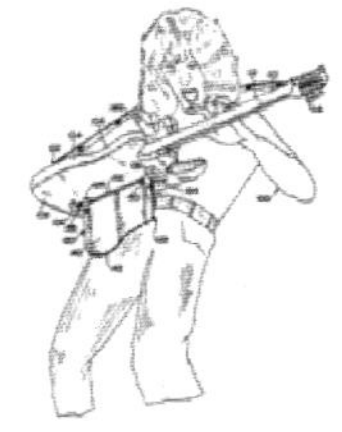

Pat. No. 4,656,917 (1987) for "Musical Instrument Support"
Edward L. Van Halen

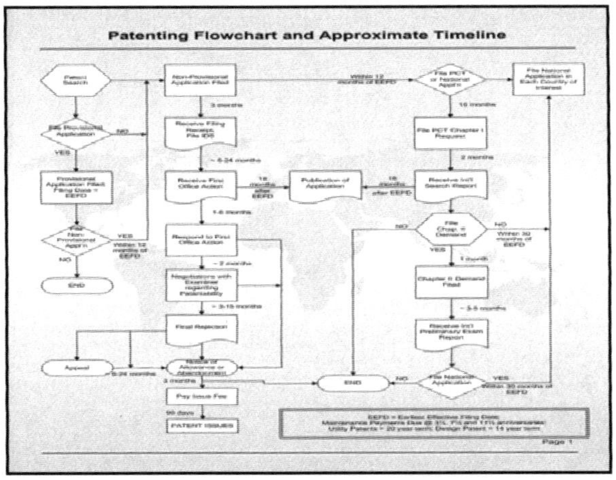

Parts of Patent Application

- Abstract (brief explanation)
- Specification
  - Field of invention
  - Background (problems solved)
  - Summary
  - Detailed description
  - Drawings
  - Claims (defines scope of monopoly)

The America Invents Act

First Inventor to File

- "First to Invent"
  - US system from 1789 to Mar. 15, 2013
  - Patent issued to party who could prove prior invention
- "First to File"
  - Most of the rest of the world
  - Patent issued to first party to file, if not anticipated by published prior art (i.e., absolute novelty).
- "First Inventor to File" – **New under the AIA**
  - US system beginning Mar. 16, 2013
  - Patent issued to first party to file, with exceptions
  - No more swearing behind prior art to show earlier "invention"
  - But 1 year grace period after inventor public disclosure

Public Disclosure Grace Period

- The AIA retains an inventor's **one year personal** grace period for filing an application after disclosure, if the public disclosure :
  - Was made by the inventor;
  - Was made by a third party that obtained the information from the inventor
- BUT, no grace period for a third party disclosure of an independently-created invention

Derivation Proceeding

- The AIA replaces the interference system with a derivation proceeding.
  - Derivation proceedings will determine whether an inventor named in an earlier-filed application derived the claimed subject matter from the inventor of a later-filed application.
  - Derivation proceedings will apply to applications having priority claims after March 15, 2013.

Scenario A

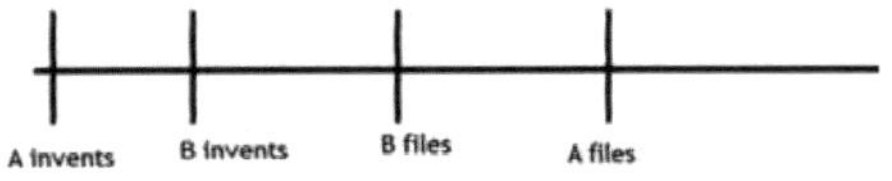

First to Invent: ***A*** obtains patent (so long as he wasn't dilatory in filing)

First to File: ***B*** obtains patent.

First Inventor to File: ***B*** obtains patent.

Scenario B

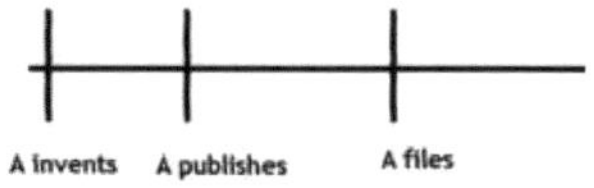

First to Invent: ***A*** obtains patent if filed within year of publication.

First to File: A does ***NOT*** obtain patent, invention in the public domain.

First Inventor to File: ***A*** obtains patent if filed within year of publication.

WHY? USA has a 1-year grace period for inventors.

Scenario C

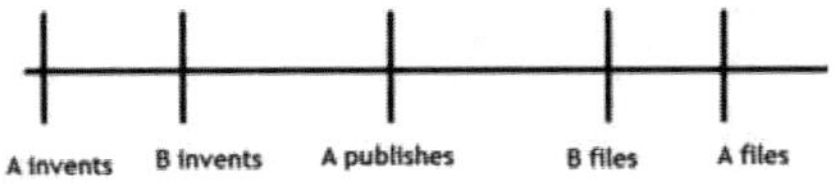

First to Invent: ***A*** obtains patent if filed within year of publication.

First to File: ***No one*** obtains patent, given to public domain.

First Inventor to File: ***A*** obtains patent if filed within year of publication.

Scenario D

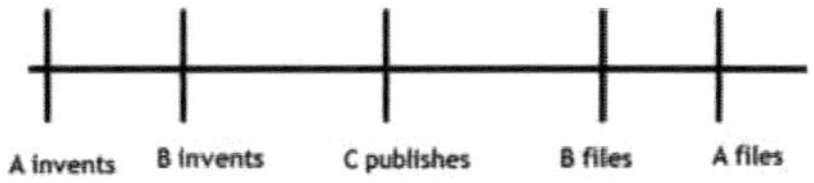

First to Invent: ***A*** obtains patent if filed within year of C's publication.

First to File: ***No one*** obtains patent, given to public domain.

First Inventor to File: ***No one*** obtains patent, given to public domain.

WHY? In new USA system, 1 year grace period limited only to publications by or derived from inventor.

Scenario E

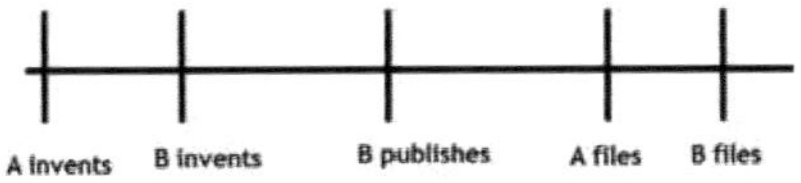

First to Invent: ***A*** obtains patent if filed within year of B's publication.

First to File: ***No one*** obtains patent, given to public domain.

First Inventor to File: ***B*** obtains the patent.

NOTE – In this scenario, because B publishes first, he receives a 1-year grace period to file, which knocks out A's filing as prior art. Thus, B is both second to invent and second to file but gets the patent anyway.

Who Benefits from the AIA?

- Companies with efficient "bench-to-office" procedures
- Companies willing to use Provisional Patent Applications or Preemptive Publication
- Universities/ research institutions and micro entities get lower fees
- Entities that may publish before filing (consider foreign effects)
- Entities willing to pay $4800 for accelerated prosecution for 6 month examination and 1 year decision

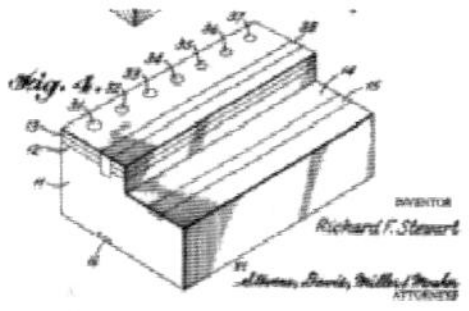

Pat. 3,138,747 (1959)
Integrated Semiconductor Chip

Owned by Texas Instruments, Inc.

Who Loses in the AIA?

- Garage Inventors
- Entities that delay filings
- Patent Applicant Pocketbooks (fees ↑)

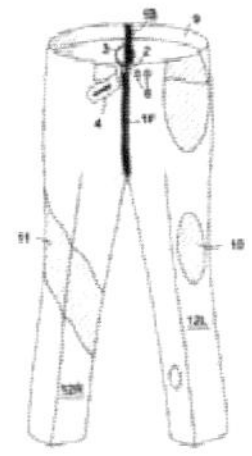

Pat. 6,161,223 (2000)

"Pants Separable at Crotch for Style Mixing"

Prior Use Defense

- Common in first-to-file countries
- Historically not available at all; added in 1999 as defense for parties making secret commercial use of certain business methods prior to patent being applied for

Pat. 223,898 (1880) for Electric Lamp
Thomas Edison

Prior User Rights

- The AIA has expanded the prior use defense to apply to *any* patent issued after Sept. 16, 2011, "consisting of a process, or consisting of a machine, manufacture, or composition of matter used in manufacturing or other commercial processes."
- Cannot be asserted against university-owned patents
- Lots of limitations

Best Mode Defense

- The AIA eliminates the best mode defense, i.e., accused infringers can no longer claim that an asserted patent is invalid due to the patent holder's failure to include the best mode.
- However, the AIA does not eliminate the best mode requirement under 35 U.S.C. § 112 for patent applicants when submitting an application.

New Post-Grant Proceedings – Effective 9/16/12

- Post-Grant Review
  - Challenge patent on any grounds within 12 months of issue
- Inter Partes Review
  - Challenge patent on novelty or obviousness grounds
  - Replaces Inter Partes Reexamination
- Supplemental Reexamination
  - Ex parte proceeding to "scrub" new prior art or inequitable conduct
  - Patent Owner initiated

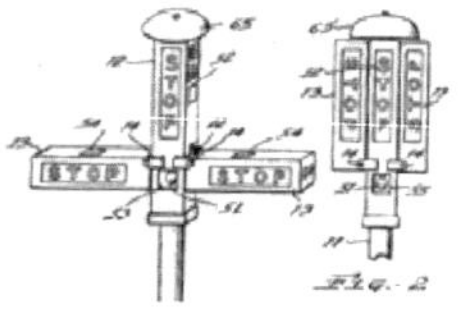

Pat. No. 1,475,024, "Traffic Signal" (1924)
G.A. Morgan

False Patent Marking Provisions

- The AIA eliminated qui tam false marking suits, but the U.S. may still sue for civil penalties.
- The provision applies to all cases filed on or after Sept. 16, 2011 **and** all pending cases.
- The AIA created a new cause of action for false marking for private parties that have suffered a competitive injury due to the false marking.
- Virtual Marking

Joinder of Multiple Defendants

- For multiple accused infringers to be joined in the same action:
  - The right to relief must be asserted against the infringers jointly and severally, or arise out of the same transaction(s) or occurrence(s) relating to the making, using, importing, offer to sell, or selling of the same accused product or process, **and**
  - Questions of fact common to all defendants will arise in the action.

Any questions?

David W. Holt
dholt@babc.com
256-517-5191

(63b) Effective IP Strategies Under the AIA

Mark Rygiel, Sterne Kessler, Washington, DC

The recently enacted America Invents Act (AIA) represents the most significant change to the U.S. Patent Law System in sixty years. The AIA effects all aspects of the U.S. patent system from obtaining patents, enforcing patents in litigation, and attacking invalid or overly broad competitor patents. This presentation will look at the AIA and how successful companies in the chemical, energy, process and product development industries can utilize effective intellectual property strategies under the AIA to protect their technology and thwart competitors. Particularly, this presentation will examine how companies can use new procedures under the AIA to quickly build a patent portfolio, strengthen issued patents, and enforce issued patents in litigation. Additionally, this presentation will address strategies for navigating the existing competitor patent landscape by using the new post-grant procedures created by the AIA.

(63c) Protecting Your Inventions Before It Is Too Late

Anthony Venturino, Novak Druce and Quigg, LLP, Washington, DC

The presentation will cover patent and trade secret basics, how to identify an invention, how to protect it by patenting or keeping it as a trade secret, how to share inventions during joint development, and how to manage employees to protect inventions and trade secrets.

(63d) Patents 201: A User-Guide to Patents in the Energy Industry

Jeffrey J. Phillips, Winston & Strawn LLP, Houston, TX

Many past AIChE presentations relating to patent law have focused on the basics of patentability. The purpose of this presentation is to help chemical engineers and their management understand how to use their patents, and how to avoid common pitfalls associated with competitors' patent portfolios. Specific topics that will be covered include:

- Patent licenses (should I grant one, do I need one from someone else);
- Enforcement of patents (how to stop an infringer);
- Obtaining defensive patents for R&D work (why the costs are usually justified);
- Freedom-to-operate opinions (do I need one, yes);
- Merger and acquisition due diligence (caveat emptor); and
- Anatomy of a "strong" patent portfolio.

This presentation is targeted for a mixed audience with technical, business, and legal backgrounds, and will include real-life examples from my experiences counseling and litigating in the petrochemical and biofuels industries.

Overview of P.E. Licensing Process in United States

William R. Parrish, P.E. and Cory D. Jensen, P.E.

Professional engineering licensure was established in the United States to protect the health, safety and welfare of the public from unscrupulous or incompetent decisions of engineers. Licensure provides a proven and accepted assessment of experience, education and ethical standards.

This presentation gives an overview of the benefits of being licensed, the key players involved in becoming licensed and the path to licensure. The overview presents the background information to help put the other session presentations in perspective. More details about licensure are available in the March issue of *Chemical Engineering Progress (CEP).*

Value of being licensed

Being licensed shows other that you have a strong commitment to high technical and ethical standards. It can provide you with better career opportunities and provide a competitive edge when seeking employment. Many employers state that they would choose a licensed engineer over an unlicensed engineer, given both are otherwise equally qualified. Also, companies often offer incentives to encourage their employees to become licensed.

Some engineers think that licensure is important only in consulting or expert testimony. However, licensure is often required in firms doing design, construction and startup of processes. State laws also may include those teaching "advanced" engineering subjects be licensed.

Some states allow an "industrial exemption" for large companies who have the financial resources to cover liabilities arising from employee decisions. Most companies encourage licensure but many oppose eliminating the industrial exemption if they have operations in multiple states because it requires their engineers to be licensed in multiple states. There is a push to eliminate the industrial exemption. This important topic is discussed in another presentation.

No one can predict the exact future course of their career. However, being licensed may provide unexpected rewards and opportunities.

Key players in licensure

Becoming a P.E. involves education, experience and examination. Besides your coworkers, who can help guide your job experience, there are two important players in becoming licensed. The first is your local engineering board that establishes the standards for education and experience and ability to take the examinations. You will work with the board throughout the licensure process.

The second important player is the National Council of Examiners for Engineering and Surveying (NCEES) (www.ncees.org) . This is a nonprofit organization whose members are the various engineering licensing boards. The NCEES was created to provide some uniformity of standards of education, experience and examination. The membership establishes Model Law to offer uniformity. However, Model Law is a set of suggested best practices and no board follows Model Law exactly. All exams are administered and graded by the NCEES.

The licensure process

Obtaining a P.E. involves the following steps:

1. Having the appropriate formal education
2. Passing the Fundamentals of Engineering (FE) exam
3. Obtaining the appropriate experience (engineering internship)
4. Providing required documentation to board
5. Passing the Chemical Professional Engineering (PE) exam

While the general requirements are uniform across all jurisdictions, exact details vary. It is critical to work with your local board so that you understand and meet the exact requirements. A brief synopsis of each step is given below. More details are provided in the *CEP* article and in other presentations in this session.

1. Formal education: All jurisdictions require BS engineering degree, or equivalent, from an EAC/ABET-accredited four-year engineering program. Foreign degrees and other degrees are considered but must be approved by the board.
   Current Model Law states that as of 2020, engineers taking the PE exam must have a Masters or equivalent (MOE). This requirement is strongly opposed by AIChE and many other organizations. More details are presented in another presentation in this session and in the *CEP* article.

2. FE Exam: This exam tests your fundamental knowledge of engineering. It is easiest to pass when you are first eligible to take it (usually senior year or shortly after graduation). Another presentation and the *CEP* article provide more details of the exam and the upcoming changes in the exam.

3. Practical experience: Your work experience must be appropriate engineering practice. Typically four years are required although some boards permit fewer. In some jurisdictions advanced degrees can reduce the number of years required.

4. Documentation: The exact format of documentation and what is required varies between jurisdictions so it critical to work with your local board on this. However, all will require proof of education, passing the FE exam and work experience. The work experience will be supported by several references, who will also vouch for your ethical behavior.

5. PE Exam: This is the final step in becoming a licensed professional engineer. The exam is designed to be passed by a "minimally competent engineer with four years of practical experience". This exam tests the practical application of your engineering knowledge. Both a session presentation and the *CEP* article provide more details on the exam.

Summary

The AIChE recommends that its members become licensed to help ensure a competent and ethical chemical engineering workforce. While the process may seem daunting at first, working with your fellow employees and your local board will make the effort much easier. Each year about 400 chemical engineers become P.E's. They have made a significant investment in themselves and possibly in their future career.

(78b) Licensure--An Academic Perspective

S. Ranil Wickramasinghe, Department of Chemical Engineering, University of Arkansas, Fayetteville, AR

Most Chemical Engineering faculty members are not licensed. Why? Is it important for faculty to be licensed? Most undergraduates take the Fundamentals of Engineering exam but do not take the Professional Engineering exam. Why? Should faculty be licensed? Is licensure important for faculty who are conducting research? Should we encourage faculty to obtain their PE license? In this presentation these questions will be discussed.

TESTING FOR COMPETENCY - THE PROFESSIONAL ENGINEERING EXAMINATION

Anne Bertelsmann, PhD, PE, Galveston Bay Refinery, Marathon Petroleum Company LP, Texas City, TX
And
Denise Chastain-Knight, PE, CFSE, CH2M HILL, Augusta, GA

Introduction

The decision to pursue engineering licensure is a personal one, and the road includes many hurdles to overcome. Engineers take this road for a variety of reasons, with different motivations and experiences along the way, and finish the trek with the PE Exam. The Principals and Practice of Engineering Examination (PE Exam) is a nationwide exam used to ensure that licensed PE's are "minimally competent". The exam must include test content applicable to the competency being measured, effectively differentiate between competent and not competent candidates, and provide a consistent competency measure from administration to administration. This paper will define minimal competency, discuss the purpose of the Professional Engineering Exam, exam format using the Chemical Engineering Exam as the example, development methodology, item quality metrics and offer suggestions for test taking preparation. It will also describe the function of the NCEES Exam Development Committee and their role in developing a fair, effective exam, and explain why it takes weeks to get a result for a computer scored exam.

Path to Licensure

Whether an engineer undertakes the path to licensure early or late in their career, each must successfully negotiate the same hurdles. Each step in the process has a specific goal to either develop or to measure the engineer's capabilities toward professional competency. Figure 1 on the next page illustrates the path to licensure.

The quest to licensure begins with obtaining a Chemical Engineering degree. The next step is to pass an exam that measures competency on engineering basics. This first exam is the Fundamentals of Engineering Exam (FE) – also sometimes called the EIT (Engineer in Training) Exam. The FE exam is typically taken either during the senior year in college or shortly after graduation. The types of items resemble "textbook" questions and are designed to test understanding of the engineering fundamentals as taught in a typical engineering degree program. Currently, the FE examination consists of a general morning session (common to all disciplines) and a choice of seven discipline specific afternoon modules. Starting in the beginning of 2014, the FE examination will convert to a computer based test, which is discipline specific (i.e. no general module). For more information on the FE exam content and the conversion to computer based testing, refer to the NCEES website (http://ncees.org).

Figure 1: Path to PE Licensure

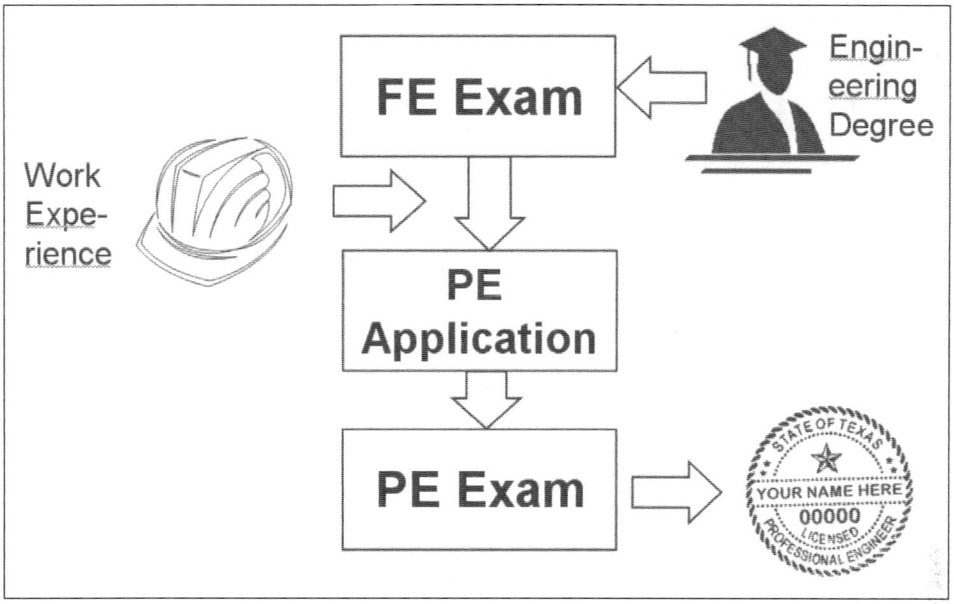

Following graduation, engineers enter the practice phase of their careers where fundamental knowledge is applied to real world application. In this phase of their careers, engineers use the theoretical principals learned from their education to solve problems that ultimately deliver physical products to market. Most states require a minimum of 4 years engineering related work experience before allowing candidates to sit for the PE exam. In order to become licensed, the candidate needs to submit an application to the state licensing board, describing education and experience, documenting passing of FE exam, and providing character and skills performance references. Format of the application and requirements for references, transcripts etc. vary from state to state. Once the application is accepted by the board, the candidate can register for the PE exam. After passing the PE examination, the engineer is licensed and can obtain the PE stamp. It is through this process that the state licensing boards ensure that only qualified candidates become licensed.

PE Exam Purpose

The PE exam is the final step in the road to licensure and its purpose is to ensure that only competent engineers become licensed. This helps to uphold the high standards of licensure and protects the health and welfare of the public. Since the exam is the same nationwide, it ensures consistency between states. Thus, one of the main requirements for a PE examination is that it can distinguish between competence and incompetence and that it can demonstrate this reliably and consistently. It is important to remember that "competent" should not be equated with "expert". The purpose of licensure is to identify individuals who have at least the minimum knowledge at which a person can practice professional engineering in such a manner that will safeguard life, health and property and promote the public welfare. That is why the term "minimally competent" is used.

A minimally competent engineer:

- should be able to perform identified tasks
- should meet minimum education and experience requirements
- is beginning to specialize in a sub-discipline
- may safely practice in another sub-discipline even though he/she does not practice in this area on a daily basis

The definition of "minimally competent" has implications on how the questions on the exam are written. The PE exam differs from the FE exam in that it addresses both the principles and the practice of engineering. So, while the FE questions are typically textbook related, questions on the PE exam are predominantly practice related. The questions build upon the experience that an engineer has gained during his or her first few years of working and need to recognize that chemical engineering experience can be gained in a variety of areas. The difficulty level of the questions is geared towards an engineer with about four years of experience.

PE Exam Content

The content of the examination is determined through a survey of practicing chemical engineers (PAKS study = Professional Activities, Knowledge and Skills), which is repeated about every 6-8 years to ensure that the examination stays current. The PAKS process includes conducting a technical content survey with input from practicing professionals. This survey is conducted over a period of weeks to ensure that the respondent pool size is adequate and is sufficiently diverse to be representative of the overall profession. Survey results are analyzed by a team of licensed Chemical Engineers, most with no prior exposure to PE exam development work, under the direction of a psychometrician who ensures that proper procedures are followed. The PAKS team will determine both the domain (subject area) content and item weight of the subdomains. PE exam content is established based on the PAKS study, and is published in the format of the PE exam specification available from the NCEES website (http://ncees.org/exams/pe-exam/) and is summarized in Table 1 on the next page.

The ChE exam specification was recently updated and became effective for the April 2011 exam administration. It includes six major domains and fifteen sub-domains that investigate theoretical and practical knowledge. Exam content is distributed within the major and subdomains according to the percentages shown in Table 1 to ensure a consistent content basis for every exam. The specification also includes guidelines for topical content within each sub-domain. For example the exam will include 4% of the items from sub-domain VI. C. Operation. Figure 2 includes example content from the 2011 exam specification.

Table 1: April 2011 ChE PE Exam Specification Summary

Knowledge Area	Approximate Percentage of Examination
I. Mass/Energy Balances and Thermodynamics	**23%**
A. Mass Balances	10%
B. Energy Balances and Thermodynamics	13%
II. Heat Transfer	**16%**
A. Mechanisms	9%
B. Applications	7%
III. Kinetics	**11%**
A. Reaction Parameters	6%
B. Reactors	5%
IV. Fluids	**16%**
A. Mechanical-Energy Balance	12%
B. Flow and Pressure Measurement Techniques	4%
V. Mass Transfer	**14%**
A. Phase Equilibria	5%
B. Continuous Vapor-Liquid Contactors	8%
C. Miscellaneous Mass Transfer Processes	1%
VI. Plant Design and Operation	**20%**
A. Economic Considerations	1%
B. Design	10%
C. Operation	4%
D. Safety, Health, and Environment	5%

Figure 2: Sample Exam Specification Sub-domain Content

C. Operation 4%

1. Process and equipment reliability (e.g., testing, preventive maintenance, startup/shutdown procedures, robustness)
2. Process improvement and troubleshooting (e.g., debottlenecking, experimental design and evaluation, optimization)

PE Exam Format:

The PE examination is given twice a year - once in April and once in October. It is an eight hour exam consisting of a four hour morning session and a four hour afternoon session with a lunch break in between. There are a total of 80 multiple choice questions (40 in the morning, 40 in the afternoon). Each question has four answer choices of which one and only one is the correct answer. Each question is stand-alone, i.e. it is independent of the other questions on the exam and can be solved with the information given in the problem, sometimes with the help of standard references. Some questions involve calculations while others are qualitative. All questions count equally and there is no penalty for guessing.

Each of the 80 multiple choice questions on the exam is of the same general format. In exam development, questions are referred to as "items" rather than questions because not every item has a question mark (like the one in the example below). Figure 3 illustrates the anatomy of an item, i.e. its general format and identifies the different parts of the item.

Figure 3: Anatomy of an Item

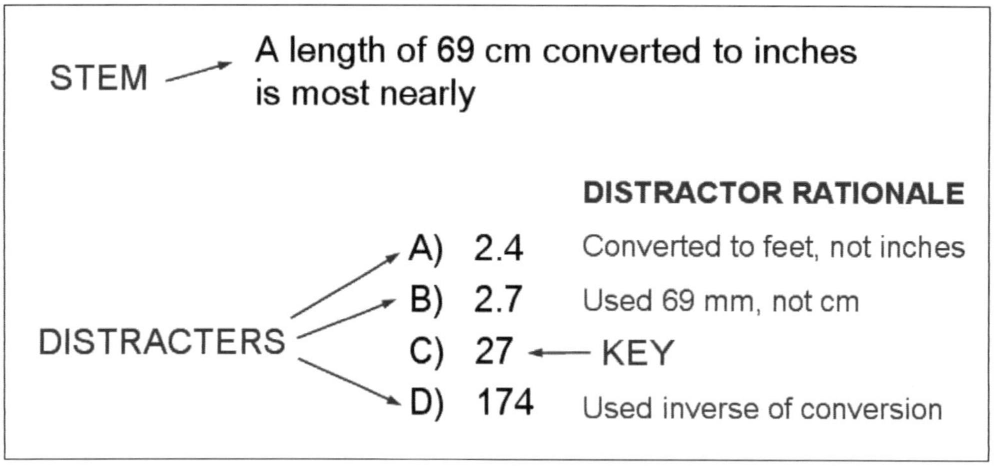

Each item consists of a stem and four answer choices. The stem (or problem statement) identifies what calculations are needed. It includes the information that is provided to solve the problem (sometimes even in table format) and states what is required for an answer. Beneath the stem are the four answer choices. Of these four answer choices, one (and only one) is the correct answer – the key. The other three are so-called distracters. They are incorrect answers – but they are not random numbers.

Each of the wrong choices represents a typical error that a less than minimally competent engineer (or someone not familiar with this specific topic) may make. Each distracter has to have a "rationale", i.e. an explanation which error was made. So, when a candidate is taking the exam and calculating a number that is exactly one of the given answer choices it does not necessarily mean that the problem was solved correctly. Maybe the candidate just made one of the "expected likely errors"…

The distracters can have a significant impact on the statistics of an item. If one of the distracters is obviously wrong (such as a negative absolute pressure), then even less than competent candidates will disregard it. That reduces the feasible choices, improves the "guessing" performance, and increases the chance that a not minimally competent candidate gets the question right. The Exam Development Committee strives to develop items that include the key, and three equally likely errors as distracters.

Exam Development

The PE Examination (as well as the FE, and the Surveying Exams) is a service that NCEES provides to the licensing boards. As such, NCEES provides the framework and the oversight for the exams, but the responsibility for the content rests with each exam committee. The committee writes the items, assembles the exams and reviews the statistics. NCEES has established rules and procedures that the exam committees follow to ensure consistency. Committees report back to NCEES with their exam quality measures. NCEES also provides the logistical support (psychometric consultant, publishing and printing, administrative support, maintaining records and the item bank, administering the exam, scoring, etc) and provides all security functions. For the larger exams, NCEES also pays and arranges for volunteer travel. For the smaller exams, this is the responsibility of the sponsoring society.

The Chemical Exam Committee is the group that actually writes and reviews the Chemical PE Exam. To be on the committee, one must be a licensed PE and express interest through the NCEES website or through a committee member. The committee has about 50 active members and meets twice a year in Clemson, South Carolina at the NCEES headquarters. Not everyone comes to every meeting, of course, so typically 30 – 35 committee members present at any given meeting. During the meeting items are written and reviewed by committee members who work with one another to critique and improve the items. Task teams also work to assemble, review and pretest exams.

Before an exam is administered, each item will have no less than six documented formal reviews and likely have been assessed by more than 20 people. Committee members are asked to complete detail oriented and often challenging engineering work and can earn PDH (professional development hours) for this activity. Developing the PE exam helps committee members to keep their engineering skills current and to give back to the profession. Without the work of this committee, the PE exam and with it the licensing wouldn't happen.

Preparing an exam that effectively and consistently measures minimal competency is a challenging task that succeeds or fails based on the ability of the committee to predict item performance. Each committee relies on a structured process and employees a variety of quality tools and techniques to improve the opportunity for success. Figure 4 is a workflow of the exam development activities. Items are written by a PE and submitted for review. Many items are prepared by the committee members at each meeting, while others are submitted by PEs who do not attend the meetings. The individuals who write and review items are referred to as Subject Matter Experts (SME). The author and two reviewers will independently work the item and develop the rationale for each distracter. Through this process the item is often revised and edited to improve its presentation and quality. This step is considered one of the most critical activities of the committee because items are the building blocks of the exam and each must meet high quality standards. Each committee member is responsible for item quality and is granted authority to modify draft items as needed to meet quality objectives, or reject items that fail criteria and cannot be edited to usable items. Finally the item is given a final check by the committee chair (or authorized designee) to verify that documentation is complete, both reviewers agree to the item content, and all quality checks have been completed, and approves the item for acceptance into the item bank. The item bank is a repository for items that are ready to be placed on exams.

Figure 4: Exam Development Process

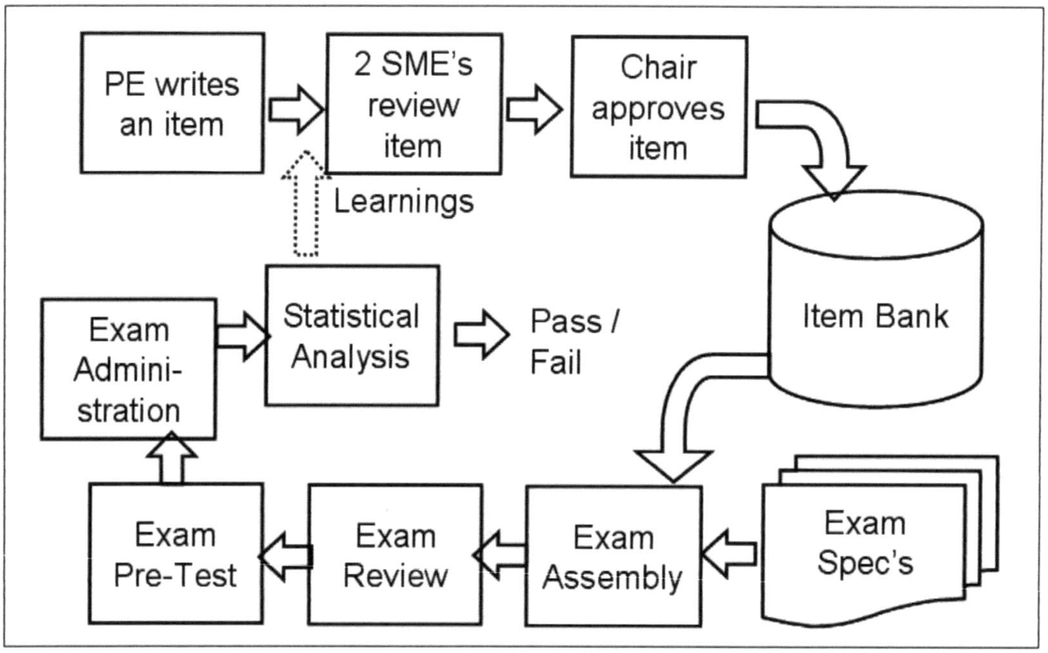

Exams are built, reviewed, revised, reviewed, quality tested and reviewed over a series of meetings. Each exam will have three independent team reviews, a pretest and a quality assessment. There are specific checks required at each workflow gate with task teams assigned to maximize the number of eyes that evaluate any given exam. An exam is compiled using items from the bank following the content distribution of the specification and sequenced in the specification order. The build team will then conduct a complete review of the exam, looking at each exam item for individual quality and checking for overall content flow and replacing items if necessary. The exam is then reviewed in two more meetings by independent teams who provide a "fresh set of eyes" and continue to improve the content to ensure that each item on the exam and the overall exam meet quality criteria. They also verify that the documentation is complete in accordance with the NCEES exam development requirements. Each exam must also have a pre-test where a group of committee members take the exam in as if they were actual test takers. After working all the items on the exam, they answer a series of questions about the items designed to predict item performance, then perform a distracter evaluation. Comments from the pre-testers are incorporated into the exam. In the final meeting before administration, a quality team conducts one last review of the exam verifying that all preceding checks (and associated documentation) are complete, all comments have been addressed, specification distribution is correct, and that author diversity limits and other NCEES criteria are met.

When the exam is administered and scored, data is compiled that can be used to evaluate the quality of the exam as a whole, and individual items. The data is analyzed and used to help the committee to learn what makes items perform effectively, and improve our performance prediction skills. Authors are provided a 'report card' so they know how items they wrote performed and have the opportunity to see how those items evolved in the review process. At each meeting, the committee will review items that had great statistics and try to find out what made them perform well. Similarly, we will review items that didn't perform well and identify traits that are undesirable in future items.

Item Statistics

Item quality is ultimately measured by the statistics generated from the candidate responses after an exam administration. Before discussing the statistical analysis of the examinee responses, one must first consider what makes an item a "good" item. The definition of a good item is one that is capable of distinguishing between "minimally competent" and not "minimally competent" candidates, does not discriminate against any particular group of test takers, and performs consistently in multiple exam administrations. To achieve those goals, it must be relevant, not confusing, and measure competency.

Most importantly, the content of the item must be relevant. Relevance is defined as the appropriateness of the topic and the difficulty of the item. The topic of the item must test knowledge or skills that are directly related to the exam specifications. For example: a question on high voltage power distribution may be relevant for an electrical engineer, but will not provide information on engineering competency of a *chemical* engineering candidate – no matter how well written it is. If the item does not relate to the exam specifications, it does not belong on the exam. Second, the difficulty level and complexity of the item must be such that a "minimally competent" engineer with 4 years experience can solve it in about 6 minutes. When writing items, authors often think about the interesting and challenging problems that are encountered in practice; however, if an engineer with 20 years experience requires several hours or days to solve the problem, what are the chances that a junior engineer will get the answer in 6 minutes? These type of problems, as interesting as they may be, do not make good exam items.

Another important principle to consider is that items should not be confusing to people who know the answer. When an item is confusing, a competent engineer who understands the concept being tested may not select the key, thus the question does not help to distinguish between competent and not competent. The so called "trick" question does not evaluate a candidate's competence and is considered a "bad" item to be weeded out in the exam development process.

It is equally undesirable to use items that give the answer away to people who don't know the answer. If the item includes clues to the key, then an incompetent engineer can choose the right answer without understanding the subject matter; thus the item does not measure competency. Clues can take many different forms:

- If some of the answer choices can be ruled out as impossible (like negative absolute pressures), the chance of guessing the right answer goes up.
- Grammar can give clues when one of the answer choices "stands out", i.e. is more detailed or does / does not match the grammatical structure of the item statement

For example, an item that asks about the 19th president of the United States, where three of the four answer choices are well known pop stars, does not require the candidate to know anything about US history to pick the correct answer.

Following an exam administration, a psychometric analysis is performed (psycho – metric = measuring the mind) on examinee responses. The analysis is solely based on information provided on the marked answer sheets without regard to the actual content of each item, but quite a bit of information about an item can be observed just from examinee responses. The psychometric analysis will evaluate a number of things. Of course, the simplest analysis is to determine how many candidates selected the key. Statistically, there is a 25% chance of guessing the correct answer because there are four choices. An item where less than 25% of the candidates selected the key may be flawed because it may have directed candidates towards preferentially selecting a distracter over the key. The next question is to ask <u>*who*</u> got the question right. Ideally, the candidates who got the question right were the same people who performed well on the exam overall. The analysis looks at the distribution of answers between the distracters and who selected them. This analysis also considers if all distracters were selected equally or if there was one distracter that was particularly attractive; maybe it even got selected more frequently than the key.

The objective of this analysis is to confirm the validity of each item as a tool capable of measuring engineering competence. An item that follows the basic three criteria (relevance, no confusion and no clues) can be expected to distinguish between competence and incompetence; however, sometimes clues, confusion or complexity can creep into an item unintentionally and be missed during the review before the exam is given. Statistics are then used to identify potential flaws. One such flaw is a mis-key. Mis-key is when the committee identified for example option A as the correct right answer, but option B is actually correct. Another flaw is an ambiguous item where more than one of the options could be correct. Despite the committee's efforts, such flawed items occasionally make it through the review process and are placed on an exam. The rigor of review process prior to exam administration minimizes the likelihood of a flawed item; however, it cannot totally eliminate them. The statistics after the exam will reflect item imperfections and provide an opportunity to compensate for item flaws in exam scoring so that candidates are not negatively impacted by flawed items.

Figure 5 is an example of the statistical analysis report for an item prepared by the psychometric consultant. One such sheet is prepared for each of the 80 items on an exam. The top left corner shows the "number of cases", i.e. the number of examinees who took the exam in this administration. In this example, there were 2910 candidates. So, this example is not from any Chemical PE exam where typically about 300 candidates sit per exam. The number below the case number (highlighted in yellow) is the Mean (or the P+ value). This number shows what percentage of the candidates selected the key; in this case, 65%. P+ alone indicates that the question is of appropriate difficulty for the average test taker; not too easy and not too hard. Statistically, the higher the P+, the easier the item.

Figure 5: Item Statistics

Number of Cases	2910
Mean (P+)	0.65
Delta	11.44
Pearson	0.33
Biserial	0.43
Mean of Passers	0.78
Mean of Failers	0.53
Mean Near Cut	0.68
Phi	0.26
DIF Grade	

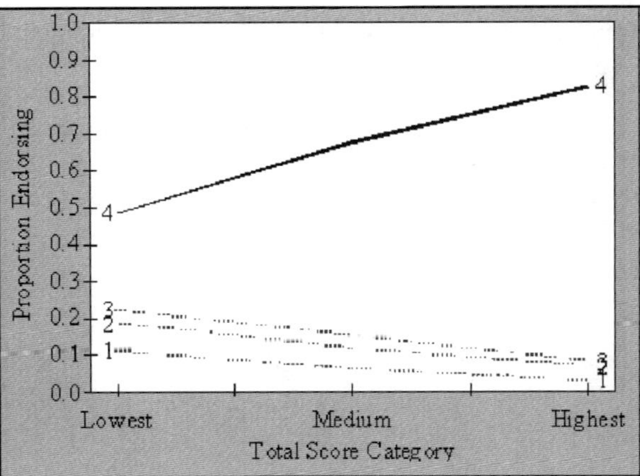

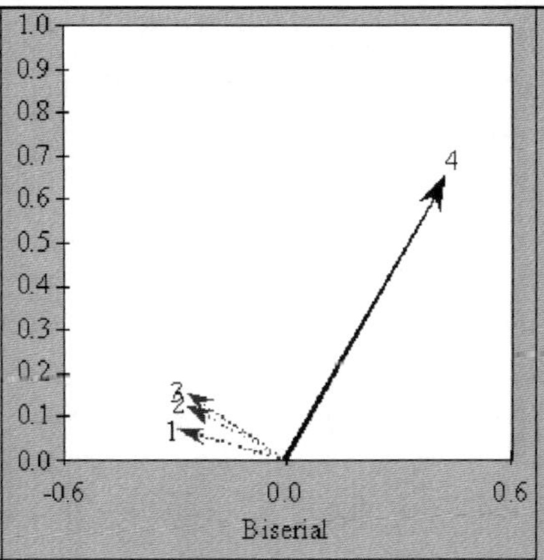

Alt.	Lowest	Low	Medium	High	Highest	%	N	Mean Scr	Biserial
1	0.11		0.06		0.03	6.7%	195	43.0	-0.29
2	0.19		0.12		0.07	12.5%	365	44.1	-0.27
3	0.22		0.15		0.08	15.4%	449	44.3	-0.27
4*	0.48		0.67		0.83	65.2%	1896	50.1	0.43
OM	0.00		0.00		0.00	0.2%	5	42.4	-0.20
Mn Scr	38.68		48.50		58.27				
Grp N	1025		1022		863				

The statistics not only evaluate how many candidates selected the key, but how selecting the key on this item correlates to overall exam scores. Test takers are divided into three groups of roughly equal size based on overall score on the exam; the low scoring third, the medium scoring third and the high scoring third. For each of these groups, the report in the table under the left graph shows how many selected which answer choice. For example on this question, "4" (or "D") was the key as indicated by the asterisk. Of the low scoring third, only 48% got the right answer, while 83% of the high performers selected the key. Thus, someone who performs well on the exam overall (which is an indication of competence) is significantly more likely to select the key than someone who scored low (may indicate lack of minimal competence). This spread shows that the question is capable of testing competency.

Figure 5 also illustrates that the incorrect answers were relatively evenly distributed between the distracters, indicating that none of the distracters appears to be overly attractive or easily ruled out. Another way of looking at the same data is presented on the right of the page. The table shows average overall scores for the people who selected each of the answer choices. In this case, the people who selected the key had an average overall score of 50, while the people who selected the distracters had overall scores in the range of 43 to 44. Five candidates with an average score of 42 did not answer the item. This clearly demonstrates that the people who selected the key were more likely to do well on the exam overall, and indicates that the item can effectively measure competency.

The data is evaluated in yet another way by the cumulative statistics in the lower left: Mean of Passers indicates that of those candidates who passed the exam, 78% selected the key. Mean of Failers shows that only 53% of the people who failed the exam selected the key. The larger the spread between these two numbers, the better the item is at measuring competency.

A handy way of summarizing the "who got the question right" analysis is the R-Biserial number. Basically, this number is related to the slope of the lines in the graph on the left. For a good item, this slope for the key should be positive and the larger the better because it expresses the difference in performance of the low and the high scorers. Theoretical range for the R-Biserial number is from -1 to +1, positive values indicate that the candidates who selected the key also scored higher on the overall exam. A negative R-Biserial number indicates that higher scoring candidates selected one of the distracters, which could be an indication of a mis-key or ambiguous item. The R-Biserial number is highlighted in yellow on the left and shown in the picture on the right. The R-Biserial number is important because it evaluates how well an item meets the three basic criteria. We previously observed that the sample item is of appropriate difficulty based on the percentage of correct answers (P+ value). The R-Biserial number shows that we did not confuse the competent candidates (a high percentage of high performers got it right) and apparently it did not give any clues to the less competent candidates (. It also appears to be relevant with regard to the topic since the high performers (those with better understanding of chemical engineering) did better on this than the low performers.

To further illustrate the concept of the R-Biserial number, consider an example of an item that might produce undesirable statistics. The item is asked on the final exam for a graduating class of chemical engineers: "Which bar in town has the most beers on tap?" One would expect a certain percentage of students to select the correct answer, but are those candidates the most knowledgeable students who spend a lot of time in the lab, or the "party crowd"? This question would likely have a negative R-Biserial number, indicating the overall least knowledgeable students are more likely to select the key, which means this item is not an effective measure of a graduating chemical engineer competency, nor is it relevant to chemical engineering knowledge and skills.

Determining Competency

The individual item statistics are the building blocks for determining the overall exam performance. To make the move from the individual item statistics to determining competency of the candidate, a passing score, also called cut score, needs to be established. Figure 6 on the next page illustrates the philosophy behind relating the score distribution to the cut point, or passing score. Total score is simply the number of items answered correctly. For the Chemical Engineering PE Exam, the total score has a theoretical range of 0 to 80. Most candidates have at least some knowledge and get lucky guessing on some questions, so scores on the very low end are uncommon.

If it were known upfront which candidates are competent and which ones are not, the scores of the two groups could be plotted as shown in Figure 6. The at least minimally competent engineers have higher scores, the not minimally competent engineers have lower scores. However, there will be some overlap between the two groups. The better individual items differentiate between competency and incompetency, the smaller the overlap. However, because there is a range of competency and because a person's performance can vary, there will always be some overlap and thus the cut point decision is not clear cut.

Figure 6: Score Distribution

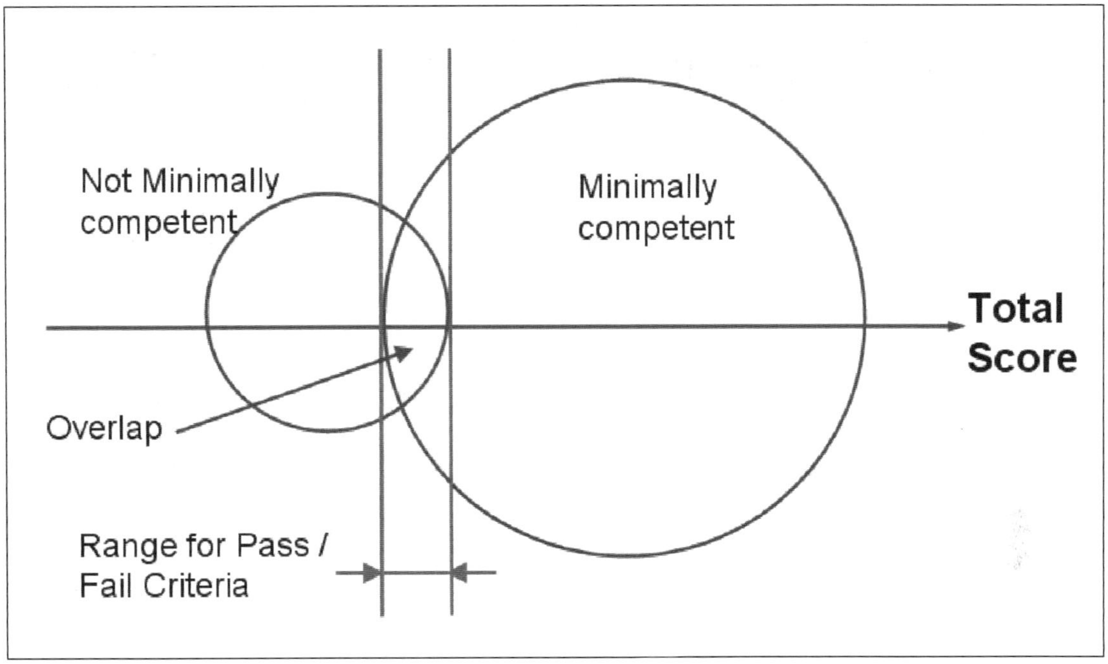

The overlap represents the range where the pass/fail decision should be made. All the at least minimally competent candidates on the right side should pass, all the not minimally competent candidates on the left should fail. With any decision point within that overlap range, there is a potential for competent engineers to fail and/or for less than minimally competent engineers to pass. Failing a competent engineer means that they will need to re-take the exam and includes the risk that they may give up on licensure. Passing a less than minimally competent engineer may allow an under-qualified person into public practice. Since the second type of error is more consequential, there is a tendency towards raising the bar when determining the pass/fail criteria.

Without knowing upfront which candidates are "minimally competent' and which ones are not (after all, it is the purpose of this exam to establish this), statistical measure are used to determine the pass/fail criteria, i.e. establish the "cut score". The cut score is the number of correct answers that a candidate needs to have in order to be considered "minimally competent". One way of determining the cut score is through panel of licensed PE's. This process is fairly involved, since it requires about 15-20 PE's who can volunteer 2-3 days of their time and travel to attend to cut score session. The panel must have good diversity (age, gender, area of practice, geographic location, demographics, etc…) and the cut score study must be completed before the results of the exam can be reported back to the state boards and ultimately to the candidates who are eagerly waiting for the scores. For these reasons, a cut score panel is usually only done when there was a change in exam specifications where the exam results cannot be directly compared to recent prior exams. The first exam where the format or specification is changed is called the "anchor exam" and cut point is established by a formal study.

For a cut score study the diverse panel of volunteers (all licensed PE's) meet and spend about a day to work the exam in a timed environment just like the actual administration. They can use any references, cannot consult with one another and they don't have access to the exam solution. The next day, they receive their score, and review and discuss each item as a group. They look at the statistics similar to those described above, but lacking the pass/fail metrics, review the solution, the distracters and discuss the topic and difficulty of each item. On the basis of that discussion, the panel members make a judgment call on how many "minimally competent" engineers out of 100 would answer this question correctly. A psychometrician guides the session and collects these estimates from each panel member. This data is then used in statistical analysis, generating a proposed passing score. The proposed passing score can then be adjusted (for example, to err on the "safe side"). This then becomes the basis for the pass/fail decision for the anchor exam.

Such a cut score study is not practical for every exam at every administration. So as long as there is no change in the exam specification, it is possible to "equate" the current exam to a previous exam. The equating process, heavily supported by a psychometrician, compares examinee performance on the current exam with examinee performance on previous exams. From this comparison it can be determined if the current exam is harder or easier than the previous exams, and cut point adjusted accordingly. If the data indicates that the exam is easier, the cut score is adjusted upward and a candidate will need to answer more questions correctly in order to pass. Conversely, candidates are not penalized if they take a harder exam because the cut point is revised downward and fewer questions must be answered correctly to pass. Equating is NOT the same as grading on a curve. When grading on a curve, the passing percentage would be fixed and the cut score adjusted to achieve that percentage. With equating, the competency of a candidate is measured on personal performance, and not dependent on the individual candidate's position within the test taking population, nor on the difficulty of the exam.

The psychometricians calculate other measures from the data. Coefficient Alpha measures the internal consistency or statistical homogeneity of the data, thus basically expresses the reliability of the exam. A measure called "Subkoviak" provides an estimate of the consistency of the pass/fail decision or an estimate of the probability that an individual would receive the same pass/fail decision on multiple administrations of the exam. All these measures are regularly reviewed to ensure that the exams can consistently measure competency.

After an exam is administered and all the answer sheets are scored, the statistical item analysis is performed. Exam Committees need to review potentially flawed items in case the key needs to be adjusted and the exam rescored. Then the cut point is determined, either through a cut score study with a panel or through equating. Only then can the pass / fail decision be made. State boards are notified with the results, perform state specific pass/fail assessments and then notify candidates of the results. NCEES recommends that only pass/fail should be reported to the candidates, though some states provide the actual score (or a "scaled percentage" based on a fixed passing percentage) to candidates. Candidates who fail get a diagnostics report so that they know how they performed in the various subject areas. NCEES recommends to only report the pass/fail decision for two reasons: The cut score can vary from exam to exam, so a candidate may pass one year and another candidate taking the exam another year may fail with the same score. The other reason is that the exam is designed to make a pass/fail decision on minimum competency. In that regard, the PE exam is different from an exam like the SAT, which is designed to measure "aptitude" over a wide range of abilities. The PE exam is not designed to measure ability accurately at the far ends of the scale. The PE exam does not distinguish between the "outstandingly competent" and the merely "extraordinarily competent engineer". Similarly, it cannot distinguish between the "totally incompetent" and the "absolutely clueless". Thus, a candidate who received a score of 75 out of 80 is not necessarily a better engineer than a candidate with a score of 72 out of 80 – both are "minimally competent" and pass.

The skeptical reader asks; "Does this really work?" Can an exam such as this really measure competency as the statistical "number magic" appears to indicate? The answer is found in the historical exam pass rates. NCEES does not report an overall pass rate for the exam, but breaks it up into a pass rate for first time takers and for repeat takers. Table 2 includes pass rates for exams given under the current Chemical PE Specification. The most recent pass rates for other exams can be found on the NCEES website.

One specific item to note about the data in Table 2 is that there is a gap between the pass rate for first time takers and for repeat takers. The pass rate for first time takers is typically rather high; in the 70 – 80% range for the Chemical PE exam. This high pass rate is because the examinee pool is already a pre-selected group. In order to be admitted to the exam, a candidate had to pass the FE exam, had to prepare an application with documentation of engineering experience, and had to be approved by the state board, a process that weeds out some of the less qualified potential candidates. Also, especially in chemical engineering where there are many jobs that come under the industrial exemption and do not require licensure, the people who decide to pursue licensure are already a non-random subset of all engineering graduates. This contributes to the relatively high pass rates for first time takers. The portion of the candidate population that passes on a repeat attempt includes in candidates that are minimally competent but may have been inadequately prepare in their first exam sitting.

Table 2: Chemical PE Exam Pass Rates

Chemical PE Exam Pass Rates	April 2011	October 2011	April 2012	October 2012
1st Time Pass Rate	78%	75%	74%	71%
Repeat Pass Rate	31%	29%	31%	20%

How to Prepare for the Exam

All the fascinating information presented in this paper aside, chances are the reader really wants to know how to prepare for the exam. Preparation philosophy will differ for a young candidate near the 4 year experience mark from that for a veteran engineer. One theory is that any minimally competent engineer can pass the examination without any preparation if he or she had unlimited time - time to look up references, read up on concepts and topics outside one's familiar area of practice, find the right equations and conversion factors, etc. But there is a time limit, and 6 minutes per question is not a lot of time.

Exam preparation should focus on building speed and efficiency for solving items. Through study and preparation, a candidate should become familiar with reference content, placing tabs to help find frequently used tables or compile a personal reference of frequently used information. Common equations and frequently used conversion factors should be memorized. Candidates should reactivate the habit of using math: solving equations, working with variables and using a scientific calculator instead of a computer. Speaking of calculators, it is strongly recommended to prepare for the exam using one of the approved models. Precious time can be wasted searching for the logarithm button on an unfamiliar calculator. Approved calculator models can be found on the NCEES website.

Exam readiness is achieved through lots and lots of practice. Preparation classes are useful for refreshing material basics, but are no substitute for rehearsal. A candidate should prepare by solving as many sample problems as one can find. By solving many, many problems, one will automatically remember the frequently used equations and conversion factors, enhance math skills and remember where to find information. The process also develops test day confidence because the items encountered on the exam have been encountered in similar form during exam preparation. Candidates should conduct one or more mock exam day(s) to practice pacing and practice maintaining focus for 8 hours. NCEES publishes a sample exam for each specification that has been developed by the exam committees in the same manner as a "real" exam. One author of this paper frequently recommends to potential exam candidates that they practice with readily available materials, complete a timed trial with the NCEES exam, then continue practicing with more materials.

Finally each candidate should prepare for the test environment and stage to be at peak performance on test day. Exams are administered in finite locations within a state and often require candidates to travel. Plan a travel itinerary that will allow arrival at the test site early, rested and unstressed. Have a satisfying but not heavy breakfast and make a plan for a relaxing lunch. Consider the test room conditions, dress comfortably and in layers. Be aware of the materials that are allowed in the test hall, and prepare accordingly. Candidates will be turned away without a valid government photo ID. Cell phones and other forbidden materials may be confiscated or admission refused. The exam is worked in a printed booklet, which contains sufficient space for working the questions and pencils are provided. Candidates need to bring an approved calculator, unapproved calculators will be rejected. The exam is open book, so candidates also need to bring any references that they want to use during the exam. References must be bound (no loose paper), but can be marked with flags. Review the test day policies on the NCEES website and plan accordingly (http://ncees.org/exams/exam-day-policies/).

Closing Comments

Professional Engineer Licensure is a personal decision and requires one to undertake a dedicated and challenging journey. Candidate should embark on this journey knowing that the exam development process is designed to ensure each exam measures for minimal competency in a consistent manner, and items are statistically validated. Finally, the factors influencing exam score processing timelines are understood. The authors of this paper encourage engineers to pursue licensure. We speak for ourselves and our fellow PEs when we say that nobody ever regrets the resources and effort spent obtaining a PE, and we appreciate the industry approval and instant expert acceptance granted by those two little letters following our names. P.E. Professional Engineer.

(78d) The Current Hot Button Issues in Domestic Professional Licensing

Amos Holt, Southwest Research Institute, San Antonio, TX

Amos Holt, a past president of ASME and current VP for Environmental, Safety and Quality Systems will review the issues currently receiving the greatest attention by the PE community which includes all of the various state and territorial licensing boards, the National Council for Examiners of Engineers and Surveyors as well as the numerous professional society stakeholders and educational organizations and societies. The discussuion will touch on the Model Law and its currect sections on the Masters or Equivalent (MOE) requirements as well as Industrial Exemptions. The various positions of key players covered including the position of a consortium of professional societies called "Licensing that Works" or LTW will be summarized. This group includes both ASME and AIChE and represents about 240,000 engineers thtroughout the world. Amos will tell you what you will likely be hearing more about in the next few years and how it might affect both current PE licensees and prospective PE licensees.

The California Divide – And Why You Need To Engage Your Legislators

Diane Spencer, PE, Lawrence Livermore National Laboratory, CA
and Emmett Miller, PE, consultant and AIChE Fellow, CA

Abstract

California is crippled by a great divide that has nothing to do with tectonic plates. California's professional engineering statute gives civil engineers free reign, allows electrical and mechanical engineers to work in their disciplines so long as that does not conflict with civil engineering, and acknowledges by title only that other engineering disciplines exist.

This division presents problems for engineers whether they have a professional engineering license or not. For example, Spill Prevention Control and Countermeasure (SPCC) plans are mandated by federal regulations. Such plans are prepared by chemical engineers all over the country, but NOT in California. SPCC plans are part of the exclusive domain of civil engineers.

A fire protection engineer was once told he could not design a fire suppression system because that was the exclusive domain of mechanical engineers. A chemical engineer was once told he could not submit a Risk Management Plan (RMP) (required by Cal EPA) without a civil engineer stamp because even the hazard analysis on a facility that uses chemicals is the civil engineer's domain. These are just a few examples where engineers are not allowed to work on projects that similarly trained engineers are expected to do in the rest of the country and the rest of the world.

So how do engineers in California navigate this regulatory nightmare? And how do we influence the legislators, engineering board, and others to nudge the statute in the right direction?

California's answer included forming a coalition of professional engineering societies that work together to present a united voice. The California Legislative Council of Professional Engineers (CLCPE) not only represents the interests of engineering professions to the legislature, but serves as a pro bono resource for the legislature on technical issues as well.

This paper will discuss: (1) our lessons learned from working with legislators; (2) California's professional engineering statute and why it is written as it is; and (3) why one needs to be engaged in the legislative process. Professional engineers are entrusted with protecting the health, safety, and welfare of the public. Getting engaged has the benefit of letting legislators understand the impact of their decisions and allows engineers to serve the communities where they live, and provide a safer, better community for everyone.

California's Problem

While most other states utilize some form of a generic registration model for engineers, California's model is discipline-specific, with civil engineering held on a pedestal. The definition of "civil engineering" in the Professional Engineers Act (California Business and Professions Code Sections 6700-6799.) is so broad that it encompasses nearly the entire spectrum of engineering. This overly broad definition, coupled with a mandate that only civil engineers can do civil engineering, means there is little room for those in other engineering disciplines to legally work in their own professions. The law allows overlap from civil engineering into all other disciplines, but other disciplines cannot overlap into civil engineering. So, not only are chemical engineers impacted, the other engineering disciplines are impacted as well.

Section 6731 states the following, with parts underlined for emphasis:

> "Civil engineering embraces the following studies or activities in connection with fixed works for irrigation, drainage, waterpower, water supply, flood control, inland waterways, harbors, municipal improvements, railroads, highways, tunnels, airports and airways, purification of water, sewerage, refuse disposal, foundations, grading, framed and homogeneous structures, buildings, or bridges:
>
> (a) The economics of, the use and design of, materials of construction and the determination of their physical qualities.
>
> (b) The supervision of the construction of engineering structures.
>
> (c) The investigation of the laws, phenomena and forces of nature.
>
> (d) Appraisals or valuations.
>
> (e) The preparation or submission of designs, plans and specifications and engineering reports.
>
> (f) Coordination of the work of professional, technical, or special consultants.
>
> (g) Creation, preparation, or modification of electronic or computerized data in the performance of the activities described in subdivisions (a) through (f)."

For many years the conventional wisdom has been that the term "fixed works," as used in the statute, (see Section 6731 above) somehow limits the scope of civil engineering, leaving room for other engineers to practice their craft. In order to clarify where that boundary lies, Senator Mimi Walters asked the California Legislative Counsel Bureau for a legal opinion (2011). The findings were shocking. "Fixed works" includes not only the structures, but also all the attachments to the structure. So, if a mechanical engineer designs a pumping system or a chemical engineer designs a fractionating column, and that hardware and the hardware is secured to a foundation, those engineers have crossed into the exclusive domain of the civil engineer. Other engineers are similarly impacted.

When a reasonable person reviews this list of studies or activities, it becomes apparent that the list precludes even scientists from legally studying the phenomena and forces of nature. And in most jobs, engineers are expected to produce engineering reports or to coordinate the work of other professionals.

How Did California Get Into This PE Mess?

The perverse nature of the current Professional Engineers Act in California is a result of good intentions gone awry.

In March 1928, a dam northeast of Castaic, CA, broke and flooded the valley below, wiping out the town of Santa Paula and flooding other towns downstream. Hundreds of people died and millions of dollars' worth of property was damaged.

The Professional Engineers Act was then created to ensure that only qualified people (civil engineer PEs) could work on dams and other fixed works. The legislators made the law as inclusive as possible to make sure there were no loopholes. Later, the Legislature expanded the Act to include Chemical, Electrical, Mechanical, Petroleum, Industrial and Metallurgical Engineering as Title Act Disciplines. With the passage of time, the Board for Professional Engineers, Land Surveyors, and Geologists (BPELSG, aka "Engineering Board") was empowered by the Legislature to create Title Disciplines on its own authority, which it did, eventually increasing the count to 15 title protected disciplines (Title Act disciplines), These are defined by Board Rules (16CCR 400-476) and have minimal stature in law. Along the line, Mechanical and Electrical Engineering were converted by statute to practice protection (Practice Act discipline)[1].

This patch job to add mechanical engineering and electrical engineering to the Professional Engineers Act has been ineffective (per Legislative Counsel's review). Inclusion of Title Act disciplines constitutes an even lower tier where those engineers are not able to take responsible charge of work within their field of expertise. In summary, the field is far from level.

The problem with the Professional Engineers Act has been known for many years. The difference now is the Legislative Counsel's legal opinion that was issued in July 2011 (www.CLCPE.org). According to that opinion, companies that rely on engineering services are on notice that the only engineering discipline allowed to conduct engineering work (not just supervise) in California is civil engineering. If a project requires a chemical engineer, a dual-licensed engineer will be needed to meet the competency requirements of both a chemical engineer and the legal requirements of a civil engineer. A dual license is needed for the other engineering disciplines as well.

[1] Four of the title disciplines (Corrosion, Manufacturing, Quality and Safety) were dropped ~1999 for lack of an NCEES exam.

The Industrial Exemption

Common to most states, California has an Industrial Exemption that allows companies engaged in manufacturing to use unlicensed engineers in their businesses— except for civil engineering work. The justification for the Industrial Exemption has been that such employers are capable of and competent in evaluating their own engineering talent, that they do not need interference from the state in doing so, and that failure that will impact the public is subject to strict liability.

There are exemptions for people who perform some specific and narrow civil engineering services. Architects performing architecture and building contractors following building codes are examples. But no general exemption exists for engineers who are not civil engineers if they perform "civil engineering." The Industrial Exemption even states that it does not apply to civil engineering.

With the recent legal opinion, however, all engineering practice in connection with facilities that are geographically fixed is now regulated as civil engineering, and the Industrial Exemption is therefore meaningless.

Engaging the Legislature

In the late 1940's, engineering societies in California organized the California Legislative Council for Professional Engineers (CLCPE) in order to work together as a resource to assist the California legislature, support laws that benefit the public and profession, and help keep society members informed about legislation that might affect them. Application of the Professional Engineers Act was not a problem to engineers in the state so long as the Engineering Board and regulators maintained a liberal interpretation. Likewise, CLCPE did not intensify its effort to change the law until the literal interpretation ultimately reinforced by the Legislative Counsel ruling came into focus.

CLCPE's efforts to change the law began in the 1980s. Our members helped draft a bill and found a senator to carry it. In California, the legislator who carries the bill is the "author" and organizations that support the bill, such as CLCPE, are "sponsors" regardless of who actually drafts the bill. Drafting a bill and finding "authors" is a process CLCPE repeated about a dozen times.

Let's reframe the process we used in California in terms that may help others take advantage of our lessons learned.

1) <u>Coalition</u>. We had a consortium of engineering societies working together as a coalition. This provided more credibility in that we were able to speak with one voice instead of individual organizations promoting separate proposals. A coalition also allows a sponsoring organization such as CLCPE to represent a larger portion of a legislator's constituency. AIChE local section leaders met with a senator to promote the one bill that chemical engineers sponsored without going through CLCPE. The senator listened intently about the impact of current law, asked a few clarifying questions, and then asked about how many chemical engineers were in the state. She was not satisfied with our answer, stood up and thanked us for our time.

Creating a coalition of several engineering societies enabled us to afford a lobbyist. AIChE and local sections are a 501(c)(3) nonprofit organization, which restricts how much money can be spent on lobbying. CLCPE however, was organized as a 501(c)(6) nonprofit, which allows for lobbying, but since the societies and CLCPE are all volunteer organizations, we do not have an intense political presence, and that restricts campaigning for or donating to candidates or political parties. Unfortunately, the opposition we face in California has major campaign funding and they openly support candidates that vote as the opposition wants them to vote.

Another advantage of working within a coalition of engineering societies goes beyond promoting legislation. The conversations we had when we met allowed us to gain insight into the problems each of the different disciplines were facing and we heard personal testimony about those various issues. This perspective has been invaluable.

2) *Hiring a lobbyist*. Over its lifetime, CLCPE has successively employed three lobbyists who helped with efforts to change the Professional Engineers Act, in addition to supporting the overall mission. A good lobbyist helps focus the discussion and facilitate consensus on language for a given bill. They also know how to navigate through the legislature, explain the bill or other proposals to legislators, legislative staff, and to others, and to find legislators who will author a bill. Typically engineers have full-time jobs, and working a bill is very time-consuming. The first of CLCPE's lobbyists was an engineer who later became Executive Officer of the Engineering Board. The other two were professional lobbyists with legal backgrounds.

3) *Language for the bill*. Earlier this paper discussed drafting language for a bill and then entering that bill into the legislative process (i.e., finding an author). "Entering" is the key word here. In the late 1990s, our bill attracted attention from the Professional and Technical Consultants Association (PATCA). In our quest to expand the law to allow engineers to practice their own profession, we needed to be careful not to adversely impact scientists (e.g., chemists) and other technical professionals. After several meetings with PATCA, we were able to work out compatible language. Going forward on subsequent bills, we deliberately engaged with the chemists (PATCA) early in the process.

4) *Bill's "author"*. Over the decades we have been pursuing change, our bills have been authored by different senators and assemblymen. In several cases, the legislator had been an engineer in his/her former career. One such senator, who had been an aeronautical engineer, had commented that it took multiple disciplines to design an aircraft. A legislator with a technical or science background readily grasped the issue we were dealing with. One of the first committees to hear a bill dealing with engineering (i.e., the bills we were promoting) was either the Business and

Professions Committee (B&PC) in the Senate or the B&PC in the Assembly. When seeking an author we often approached members of these committees as well. The intent was to get early support and move the bill through the legislative process.

5) Finding support. In addition to sponsors, it helps to find other organizations that will support a bill through (1) letters, (2) testimonies, and (3) lobbying efforts. We collected binders of letters from individuals, municipalities, regulators, and corporations. These included Chevron, the Chemical Industry Council of California, and Genentech. The letters provided a lot of anecdotes on how this perverse law affected people. The volume of the letters attracted the attention of legislators.

Once a bill is scheduled for a vote, testimonies at the hearing can be very compelling. The city of San Diego, California Fire Prevention Officers Association, a division of the California Fire Chief Association, along with various individual department fire chiefs and fire marshals, individuals impacted by current law, and others have testified at legislative hearings over the years. The Dean of Engineering at University of California—Davis testified on behalf of the University of California at the hearing in January 2012, stating that the state is educating engineers who cannot work in their own professions. The dean was shocked to find out that the legislators still would not vote to support changing the current law.[2]

The third level of support is when companies/ organizations involve their lobbyists. However, we found it very difficult to convince companies that had lobbyists in Sacramento to get engaged at this higher level.

6) Other curve balls. We experienced numerous unexpected challenges. One such curve ball happened in 1997, when chemical engineers tried independently (without CLCPE) to change the law. The Engineering Board required that we provide a Sunrise Package[3]. Essentially converting engineers from Title Act (defined in by BPELSG Board Rules) to Practice Act (defined in the Professional Engineers Act) is viewed as creating a new engineering discipline. We were given one week to prepare this. Fortunately, CLCPE had prepared something similar 10 to 15 years earlier, which we were able to mine and then create the required Sunrise Package. The Engineering Board still did not support our bill. Note: Keeping a good archive of material produced in supporting legislation could have unexpected value in the future. The key is being able to find the needed archived information when it is needed. Another advantage is that many of the same people remained engaged in CLCPE for long periods.

---

[2] Incidentally, the western region of ASCE sent their lobbyist to testify in opposition at the same hearing where the UCD dean testified.

[3] Sunrise is a process under which an occupation or profession wishing to receive state certification or licensure must propose the components of the legislation, along with cost and benefit estimates of the proposed regulation

What We Proposed and Why

Our approach to changing legislation evolved over the years. As mentioned earlier, chemical engineers made an attempt in 1997 to raise chemical engineering to a Practice Act (defined by statute). Our opposition protested that our attempt was part of a turf battle among the engineering disciplines. Legislators do not want to be drawn into controversies (e.g., infighting). Our next attempt involved CLCPE, with the intent to define all the Title Act disciplines as Practice Act disciplines. Because we had a coalition involving more than 15 engineering societies, we were able to combat this accusation. Still we did not succeed.

We also shifted our approach and proposed legislation that was closer in alignment with generic engineering. Specifically, we proposed that engineers should be allowed to do work based on their education, experience, or testing. In talking to legislators and to our lobbyist, we learned that we needed to approach any change in engineering statute with baby steps. Adopting generic licensure language (similar to other states or similar to Model Law proposed by the National Society of Professional Engineers (NSPE) was too much at one time. Because of the heavy influence by civil engineering companies in the state, we drafted bills that would continue all the advantages afforded to civil engineering. Unfortunately, our opposition would not negotiate.

In 2000, CLCPE had caught the attention of the legislature sufficient enough that they commissioned a study of the two-tier approach (Title Act/Practice Act) used in California. The study considered natural overlap between the professions, whether there was enough distinction between the engineering disciplines to support this unique regulatory structure, and whether the three Practice Act disciplines (civil, mechanical, and electrical engineering) posed more of a risk to the public (relative to the other engineering disciplines) and thus justified the existing two-tier structure. (See **www.CLCPE.org** for study by the Institute for Social Research.)

The study strongly supported changing the existing regulatory structure. The task of making this change fell to the Engineering Board. The bill it sponsored was opposed by American Consulting Engineers Council (ACEC) - California (a contractors' association) and Professional Engineers in California Government (PECG, a public employee union). Both were well-funded with strong ties to legislators — and the bill failed. [PECG is allowed to donate directly to candidates and its 13,000 members individually campaign for legislators. That has a big influence on those legislators who might support our position.]

The study marked a shift in how we approach changing legislation; CLCPE did not sponsor bills after the study. Instead, CLCPE and our individual engineering societies supported a succession of bills that were sponsored by the Engineering Board or by companies and organizations that use engineering services.

The disadvantage of this approach is that we had less influence on making amendments to the language in the bills. So while sponsoring our own bills had not been successful, this other approach allowed us to continue to pursue fixing the Professional Engineers Act.

Another challenge we had was finding companies and organizations that would commit their lobbyists. Many of these companies were very willing to prepare letters of support, but were not ready to involve their lobbyists.

End Results

This paper provides numerous lessons learned from decades of experience in trying to change California law to allow all engineers to practice in their own professions without finding a civil (typically) or other Practice Act engineer to take responsible charge of the work. Despite years of effort, we have not succeeded in changing the Professional Engineers Act. We are debating our next steps. It is our hope that the lessons we learned will be helpful to others who find the need to engage their state legislators.

(86a) Explosion and Fire At Caribbean Petroleum Tank Terminal in Bayamòn, Puerto Rico

Vidisha Parasram and Maria Mazzocchi, Incident Screening, US Chemical Safety Board, Washington, DC

On October 23, 2009, normal tank filling operation became catastrophic when a 5,000,000 gallon atmospheric tank overflowed for approximately 30 minutes at the Caribbean Petroleum terminal in Bayamòn, Puerto Rico. The developing vapor cloud enveloped part of the tank farm and reached the waste water treatment area where it ignited. A flash fire and explosion ensued and the contents of the tank burned for 2.5 days. Neighboring communities were evacuated and oil contaminated the nearby wetlands. This incident revealed a number of safety system deficiencies at the facility and gaps in the regulatory requirements and industry standards addressing overfill prevention and emergency response. This incident also raises questions regarding the adequacy of safety management systems in place at tank terminals and the need for automated overfill prevention alarms including high, high high level alarm and independent alarms to detect and shut down flow into a tank during filling operations. In addition, due to the close proximity of the community to the Caribbean Petroleum tank terminal, this incident further demonstrates the need for a risk assessment process that addresses high hazard facilities. This presentation will provide an overview of the US Chemical Safety Board's Caribbean Petroleum investigation and discuss recommendations made by the CSB to prevent a similar incident.

(86b) Use Dynamic Simulation to Reduce Startup Flare Emissions for Ethylene Plants

Jian Zhang, Meiqian Wang, Jie Fu, Chuanyu Zhao, Ha Dinh, Shujing Zhang, Qiang Xu and Kuyen Li, Dan F. Smith Department of Chemical Engineering, Lamar University, Beaumont, TX

Ethylene plants play an important role in petrochemical industry. The products of an ethylene plant, such as ethylene, propylene, butadiene, benzene, toluene and other co-products, are all basic feedstocks in petrochemical industry. Ethylene production process is one of the biggest and most complicated processes in petrochemical industry. Startup of an ethylene plant is also very critical. A typical startup of an ethylene plant could flare 400,000 - 1,000,000 lbs of materials, and generate 200,000 lbs of CO and VOCs (volatile organic compounds). Reduction flare emissions will help reducing environmental impact and saving materials.

In this paper, rigorous steady-state and dynamic models of an ethylene plant was built. The models include CGC (charge gas compressor), chilling train, demethanizer, deethanizer, depropanizer, C2 splitter and C3 splitter sections. Both steady-state and dynamic models have many details so that they are more close to startup procedure. For example, performance curves were used in CGC section, so that the section can simulate scenarios on different feed and bypass flowrate. In chilling train section, information of equipment size and heat capacity is considered for simulating precooling process. For distillation columns, rigorous distillation model and tray sizing calculation are employed. Both steady-state and dynamic models are tuned and validated to make them close to real plant operations.

The validated dynamic model was used to simulate startup scenarios. A startup procedure has dozen of steps, and some steps have several flare reduction options. Through dynamic simulation, all major steps and options can be simulated and evaluated, and the flare emissions can be calculated. For example, CGC startup is the most critical step and has largest amount of flare vent, because charge gas has to be continuously produced through furnaces and sent to flare to keep inlet pressure of compressors. We simulated several scenario cases with charge gas of different flowrate, methane and nitrogen, and found the best scenario, which could largely reduce flare emissions of CGC section. Another case study is regarding the precooling of chilling train. A chilling train should not be chilled very fast, otherwise it could be damaged by thermal expansion. On the other hand, chilling train cannot work properly before chilled to certain temperature. Through dynamic simulation, several chilling scenarios were simulated. For each scenario, control strategies are employed to control the chilling rate within a certain range. Simulation results showed some scenarios could even cool down chilling train without any flare emissions.

Flare calculation showed that the total flare emissions could be greatly reduced through dynamic simulation. The final scenarios have been acknowledged by ethylene plants.

(86c) Development of Novel Reducing Agents for NOx Reduction by Sncr Method

Maoqi Feng, Div. of Chemistry and Chemical Engineering, Southwest Research Institute, San Antonio, TX and Reggie Zhan, Emissions and Vehicle Research Division, Southwest Research Institute, San Antonio, TX

Nitric oxides (NO_x), mainly including nitric oxide (NO) and nitrogen dioxide (NO_2), are of the most toxic pollutants formed during combustion and refining processes. Post-combustion treatment methods include selective catalytic reduction (SCR), selective noncatalytic reduction (SNCR), reburning, and the combination of these aftertreatment technologies. Among the post-combustion technologies, the SNCR technology is considered to be an effective approach to reduce NO_x, with reasonable capital investment and operation cost. To reduce NO_x in a lean environment, a reductant is needed. For SNCR, the most commonly used reductant is ammonia or urea water solution as an alternative to NH_3. One of the disadvantages of ammonia is the cost of the injection system. Different reducing agents for NOx removal by SNCR method will be evaluated in this paper, including carbon monoxide, hydrogen, char, hydrocarbons, and oxygenates, etc.

Rotating Zigzag Bed Industrial Application Research on Ammonia Stripping

Zhi-chao Xu, Xiao-hua Li, Guang-quan Wang, Jian-bing Ji

(College of Chemical Engineering and Materials Science, Zhejiang University of Technology, Zhejiang Province Key Laboratory of Biofuel, Hangzhou, Zhejiang, China 310032)

Abstract: About 400 units of rotating zigzag bed (RZB) have been used in industrial distillation process. These successful applications fully demonstrated its advantages such as high gas-liquid mass transfer rate and small size. There will also be broad prospects of RZB applied in waste water treatment process. In this paper a preliminary attempt was made to study ammonia nitrogen removal from industrial waste water by RZB before emission. Firstly, stripping experiments were carried out to remove ammonia nitrogen from the waste water containing ammonia and ammonium sulfate in a small single-rotor RZB, obtaining ammonia solution with a proper concentration. Secondly, a unit of industrial-scale three-rotor RZB with outer diameter of 750mm was designed based on experimental data and used to treat ammonium sulfate solution with 2% ammonia nitrogen. The industrial results showed that the feed capacity was greater than 1500 kg/hr, ammonia concentration in the bottom was less than 30 mg/L, and aqua ammonia concentration in the top was greater than 25%.

With the improvement of human life, there is growing public concern about environment problems. Wastewater emission is one of the important factors contributing to environment deterioration. In the wastewater emissions, ammonia nitrogen concentration is a very important target which should be strictly controlled because excessive ammonia in the river will cause water eutrophication. The ammonia wastewater has a wide variety of sources, such as refining, fertilizer, inorganic chemicals, pesticide, ferroalloys, glass and food industries. Therefore, studies on removal ammonia from the wastewater and how to recover ammonia to reuse is of great significance. Rotating zigzag bed (abbreviated to RZB) has been successfully applied in about 400 cases of industrial distillation[1-5]. Due to its higher mass transfer rate it was expected to have a very wide application in wastewater treatment. Some traditional enterprises have demand for retrofit and optimization of the production process without occupied space increase because of limited land area. RZB can exactly meet this demand owing to its very small volume. In this paper lab-scale experiment of stripping ammonia from wastewater with RZB was conducted and ammonia solution with a certain concentration was obtained. Then, the experimental results were scaled up to design a unit of RZB successfully used in industrial production.

1.Lab RZB device and flowsheet

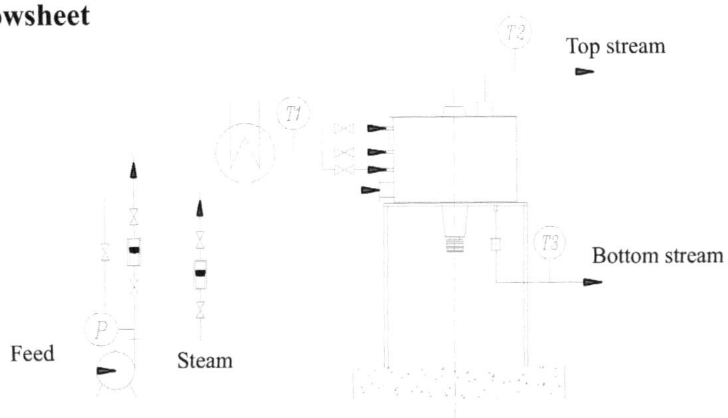

Fig. 1 Lab RZB device and flowsheet

Lab RZB device was set with multiple liquid feed inlets, which were used in variety of processes with different number of theoretical stages. For feeds with different concentration of ammonia, their optimal operation and equipment parameters can be determined by experiments.

In experimental flowsheet shown as Fig.1, the liquid feed was measured with a rotormeter and pumped into RZB device with multiple inlets after heating by a heat exchange. The steam with higher temperature enters into vapor inlet after flow measurement. The treated water was drawn from liquid outlet on the bottom of RZB. Vapor phase containing ammonia withdrawing from the top of RZB was condensed and absorbed in the next steps. T1-T3 were thermometers to measure and record the temperatures of several key positions.

2. Experiments and results analysis

In the experiments, the concentration of the prepared feed was close to that of industrial stuff, which was ammonium sulfate solution with about two percent ammonium. The feed position can be determined through the preliminary experiments. For this study, the flow ratio of feed to steam was varied and analyzed the ammonium compositions of bottom and top outlet stream as well as the power consumption of the motor. These data provided the basis for scale-up design of RZB.

Feed preparation: Ammonium sulfate solution with concentration of 15.5%, which pH was adjusted to about 13 by dripping NaOH solution, was prepared by adding ammonium sulfate powder into water. Ammonium sulfate solution has ammonium concentration of 3.3% and was chosen as the feed.

In the experiments, originally the top outlet vapor stream entered into an absorber but not a condenser. The system was maintained under vacuum condition by adjusting the relief valve at the end of absorber. The pressure of the system was 72kPa and the temperature of the top outlet vapor phase was about 80℃. The typical process and operation parameters in the experiments were shown in Table 1:

Table 1 Typical process and operation parameters

feed				steam		Bottom stream	Top stream
$(NH_4)_2SO_4$ %	Ammonium concentration %	pH	flowrate L/hr	flowrate kg/hr	pressure MPa	Ammonium concentration mg/L	Ammonium concentration(vapor) %
15.5	3.3	13	1000	120	0.35	<100	>48

If the top outlet vapor stream was connected to a condenser, solution with ammonium concentration higher than 20% can be obtained.

3. Industrial application of RZB in ammonium stripping

Fig. 2 Industrial equipment

After the results of industrial field tests were accepted by the customer, scale-up design was made for RZB. The section between the feed position and the top, corresponding to the rectifying section of distillation column, has a separation ability of 4.5 theoretical plates. This section can effectively prevent entrainment. On the other hand, if the actual ammonium concentration was lower than designed, partial reflux can be introduced to increase the ammonium concentration in the top outlet stream. The type of RZB is BZ750-3P, which means the largest one in the three rotors has a diameter of 750mm. Its appearance was shown in Figure 2.

The RZB was used in Taiwan, with industrial process flowsheet shown in Figure 3. The feed and alkali solution for adjusting the pH value, which flowrate were measured by flowmeter respectively, were mixed in a fixed proportion. The mixed stream was heated by the bottom outlet liquid and entered into the RZB. The steam was introduced by vapor inlet. The outlet vapor entered into absorption system which performs cooling and absorption steps. The absorption systems consist of four sections: pure water (absorbent), low concentrated ammonium recycle, high concentrated ammonium recycle and finished ammonium. The flow order is pure water section → low concentrated ammonia section → high concentrated ammonia section → finished ammonia section. The finished ammonia left from the bottom of the absorption system.

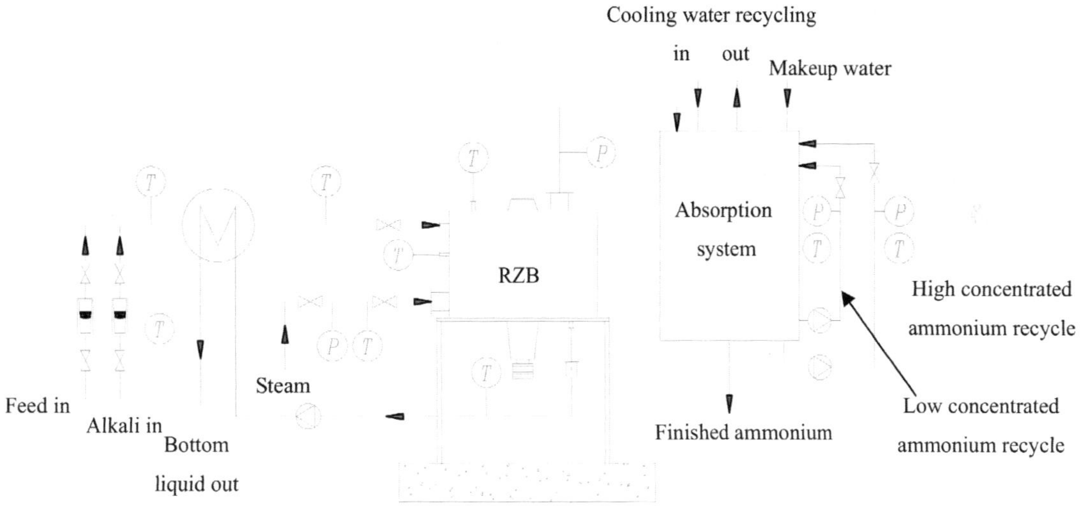

Fig. 3 flowsheet of industrial application of RZB
Ⓚ-Pump (P)-Manometer (T)-Thermometer

Actual operating parameters are as follows: the feed rate 1500kg/hr, ammonium concentration 18g/L, pH value 11.5-12, outlet vapor temperature about 80℃, no reflux, finished ammonia concentration > 25%, ammonia in outlet liquid<30mg/L, unit consumption of steam 90kg/ton feed, bottom temperature about 98℃. the temperature approach between inlet steam and heated feed was about 10℃. The customer was very satisfied with the results.

4. Conclusion

About 400 units of RZB used in a variety of industrial distillation processes demonstrated its higher separation ability. Its successful application in stripping ammonium from the wastewater was another good example.

RZB has a small volume with a height of lower than 1m. For traditional enterprises with small unused space in the factory, this kind of device can find its place here.

In this system, feed was heated by bottom outlet stream which sensible heat was recovered. So the system shows energy-saving effect.

5. Prospects

Modification of existing RZB device can realize liquid cut off between its rotors, which can extend its applications in recovery of ammonium from the wastewater. For RZB with more than three rotors, its top and bottom section perform absorption and stripping respectively, thus these two processes can be integrated in single equipment.

Reference

1 Ramshaw, C. (1983)., "Higee distillation—An example of process intensification, " Chem. Eng., February, pp. 13–14

2 Rao, D.P., Bhowal, A. and Goswami, P.S. (2004), "Process intensification inrotating packed beds(HIGEE): An appraisal," Ind. Eng. Chem. Res., pp. 1150-1162.

3 Li, X.P., Liu, Y.Z. and Li, Z.Q. (2008), "Continues distillation experiment with rotating packed bed," Chin. J. Chem. Eng., pp. 656-662.

4 Wang, G.Q., Xu, Z.C., Yu, Y.L.and Ji, J.B. (2008), "Performance of a rotating zigzag bed—A new HIGEE," Chem Eng Process, November, pp. 2131–2139.

5 Ji, J.B. , etc. (2004), "Equipment of zigzag high-gravity rotating beds," CN Pat., 01134321.4.

(106a) Challenges and Opportunities in Oil Fields Produced Water Treatment

Rafique Janjua, Process, Fluor Enterprise Inc, Sugar Land, TX

Produced water treatment composition varies all over the map and requires specific unit processes and unit operations with each and every site. Produced water from Canadian Oil Sands is codcndate from high pressure steam flooging/injection into Bitumen layer It have heavy oil emulsion, high Silica and dissolved solids as compared to tradional oil fields with more free oils, very high salinity and heavy metals. Produced or surface water used for Enhanced Oil Recovery have differnt reqiurements, such as suspended solids removal, corrosion and scale protection, complete Oxygen scavanging, selective Sulfate removal and Biocide injection which will require totally different treatment process. There are different challanges and opportunities for cost effective and reliable sytems for each application. This paper will discuss some real life examples how to achieve the desired results.

(106c) Sustainability in Industrial Water Treatment System

Vina Arjomandnia, Process Technology, Fluor, Sugar Land, TX

Sustainability in industrial water treatment system

Water is as important in the oil industry as the crude feed itself.

Proper water treatment is essential for reliable and efficient refinery operation as well as compliance with social and environmental regulations.

Sustainable water treatment system in industry should provide adequate water quantity and quality for general and specific needs such as high pressure steam generation. Wastewater streams selective source segregation and treatment enhances maximum sustainability and recovery for reuse as make up water in cooling towers etc. Non contaminated rain water along with treated contaminated rain runoff can provide a sustainable water supply for industry.

This paper will discuss designs where these methods could be implemented to ensure the sustainability of the industrial water demands, on domestic and international projects.

(106d) Overview of Boiler Water Treatment and Associated Wastes

William Moore, Industrial, WesTech Engineering, Salt Lake City, UT

Overview of Boiler Water Treatment and Associated Wastes

Boiler water treatment consists of steps to remove contaminants that will corrode or cause scale on the boiler tubes. As the boiler pressure increases, more contaminants must be removed. There are also various chemicals added to the boiler water. These chemicals and the wastes from the water treatment systems can produce an interesting soup for discharge. This paper discusses the systems used to produce boiler feed water and the ensuing discharge chemistry.

Genetic Algorithm Approach for Optimal Control of Air Emissions from an Oil Refinery

Mohammed S. Ba-Shammakh *
Chemical Engineering Dept., King Fahd University of Petroleum and Minerals,
P.O Box 5050 (31261), Dhahran, Saudi Arabia

E-mail: shammakh@kfupm.edu.sa

Abstract

Air pollution is a major issue facing the world nowadays. Many human activities and several industries have contributed to this problem. Oil refinery is an example of that and different air emissions are emitted. Examples of these emissions are; SO_2, NO_x and CO. Various studies have been conducted for air pollution control from previous decades, but no researches were extended about selecting air emissions control methods for oil refinery units considering, simultaneously, cost and efficiency of the control options. The aim of this study is to determine suitable control technique for oil refinery with two objective functions including minimization of cost and air emissions according to environmental standards. Genetic Algorithm approach is used to select an optimal control technology for each air emissions considered. The model is applied into a case study to show its application.

Key words: Air emissions, Genetic algorithm, Oil refinery

Introduction

The refining industry plays a very important role in international economics and in our daily life. Environmental regulations are pressing the refinery industry to minimize its air emissions. Petroleum refining converts crude oil into useable products, such as gasoline, asphalt, and jet fuel. This process heats crude oil to separate its components into several final products. By altering the temperature and the specific gravity of the crude oil, refineries may alter the overall composition of their final products.

The petroleum refining process results in the release of a number of air emissions from different sources or processes, including but not limited to SO_2, NO_x, and CO. In addition to the previous mentioned emissions, many greenhouse gasses (GHGs) are released from this industry but will not be the focus of this paper.

Sulfur dioxide (SO_2) itself can cause adverse effects on respiratory systems of humans and animals, and damage to vegetation. When dissolved by water vapor in air to form acids (acid gas) it can again have adverse effects on the respiratory systems of humans and animals, and it can cause damage to vegetation, buildings and materials, and contribute to acidification of aquatic and terrestrial ecosystems. When transformed into sulphate particles that are subsequently deposited on aquatic and terrestrial ecosystems, acidification can result, and when sulphate is combined with other compounds in the atmosphere, such as ammonia, it becomes an important contributor to the secondary formation of particulate matters (PM). Main sources of the SO_2 emissions in a petroleum refinery include combustion sources, thermal or catalytic cracking processes, reforming process and sulfur recovery plant.

Nitrogen oxide (NO_x) is another main air pollutant from a petroleum refining and can be defined as a generic term for the various nitrogen oxides produced during combustion process. They are believed to aggravate asthmatic conditions, react with the oxygen in the air to produce ozone, which is also an irritant and eventually form nitric acid when dissolved in water. The main sources of this pollutant in a petroleum refining are combustion processes.
Another major pollutant from a petroleum refining is carbon monoxide (CO). It is a colorless, odorless, tasteless and poisonous gas. It is a product of incomplete combustion of hydrocarbon-based fuels. CO can have a significant impact on human health. It enters the bloodstream through the lungs and forms carboxyl hemoglobin, a compound that inhibits the blood's capacity to carry oxygen to organs and tissues. The main source of this pollutant in a refinery is combustion process.

Since combustion is the major source for SO_2 emissions in an oil refinery, most of the research dealing with SO_2 control technologies, in an oil refinery, focused on either reducing the output from a unit that emits more SO_2 or control the emissions after it is emitted. These may include switching to run with fuel that has low sulfur contents or to control SO_2 emissions by installation of different control technologies. Each of these technologies has its own cost and it will defiantly affect the profit of a given refinery if it is installed or implemented.

Emissions control measures leading to reductions in NO_x is expected to reduce population exposures to all gaseous NO_x. Many technologies have been invented for NO_X reduction. Selective Catalytic Reduction (SCR), Selective Non-catalytic Reduction (SNCR) and low NOx burner are common technologies in the USA for oil refinery NO_X reduction.

Table 1 lists some common technologies that can be used for each pollutant under consideration.

Table 1: Air emissions with common technologies

Technology	NO_x	SO_2	CO
SNCR	X	-	-
SCR	X	-	-
Low-NO_x Burner	X	-	-
Thermal incinerator	-	-	X
Catalytic incinerator	-	-	X
Wet Scrubber	-	X	-
Spray Dry Scrubber	-	X	-

Many studies have been conducted in the field of air pollution control considering cost and efficiency [1-3]. The issue is very effective on point sources pollution control especially stacks and efforts have been conducted by several researchers [3-6]. To select effective control options with least cost, optimization approach should be used meeting air emissions standards. Optimization methods, heuristic or stochastic, are inherent for choosing suitable control technologies for different air emissions within an oil refinery [7].

Talking about modeling approach, the current available dispersion models are either analytical or stochastic. Analytical models, such as Box model, Plume Gaussian and Multiple-cell model, are based on mass transfer in air [8] whereas stochastic models such as Artificial Neural Network (ANN) or Fuzzy Logic (FL) are based on trained data [9]. Dispersion depends on the weather condition and how much air is needed to transfer and dilute pollutants. Dispersion of pollutants in air depends on different factors including wind velocity, wind direction and ambient air temperature. Multiple-cell model was used for displaying these factors.

The mathematical description of these processes leads to a system of partial differential equations as follows:

$$\frac{\partial C}{\partial t} = -\frac{\partial(uC)}{\partial x} - \frac{\partial(vC)}{\partial y} - \frac{\partial(wC)}{\partial z} + \frac{\partial}{\partial x}(K_x \frac{\partial C}{\partial x}) + \frac{\partial}{\partial y}(K_y \frac{\partial C}{\partial y}) + \frac{\partial}{\partial z}(K_z \frac{\partial C}{\partial z}) + E_s - (k_1^S + k_2^S)C + Q_s(c_1, c_2, \ldots, c_q),\ s = 1,2,\ldots,q \qquad (1)$$

C : Concentration of the air emissions involved in model

U : Wind velocity,

K_x, K_y and K_z : diffusion coefficients,

E_s : Emission source,

k_1^s and k_2^s Deposition coefficients (for the dry deposition and wet deposition respectively)

Q_s for chemical reaction.

The assumptions are:

1. Steady state
2. $U_y=U_z=0$
3. ($k_x=0$) since transport by bulk movement in the x-direction surpasses diffusion
4. No reaction (Q=0)
5. No deposition in the system ($k_1^s = k_2^s = 0$)

By considering the assumptions, equation (1) reduces to:

$$\frac{\partial(uC)}{\partial x} = \frac{\partial}{\partial y}(K_y \frac{\partial C}{\partial y}) + \frac{\partial}{\partial z}(K_z \frac{\partial C}{\partial z}) + E_s \qquad (2)$$

The finite difference method is employed for solving the above equation and the following boundary and initial conditions are used:

at x =0, $C(0,j,k)=0$

at y =0, $\frac{\partial C}{\partial y}=0$

at y =W, $\frac{\partial C}{\partial y}=0$

at z =0, $\frac{\partial C}{\partial z}=0$

$\frac{\partial C}{\partial z}=0$ at z = mixing height ,

After determining boundary and initial conditions, Eddy diffusivity, wind velocity profile and others are taken from literature [10].

Genetic Algorithm Approach

Genetic Algorithm (GA) was developed by John Holland in the 1960s and it is a stochastic optimization method for discrete variables [11]. The algorithm starts with randomly generating population of chromosomes and utilizing genetic operators that include selection, crossover and mutation operators [12]. The selection operator chooses chromosomes from the current population to produce new population. The crossover operator makes two new chromosomes from two existing chromosomes by cutting them at a random position and exchanging the parts of the following cuts. The mutation changes the chromosome genes randomly. These operators simultaneously guide random search method, which can yield an optimum set of function variables.

Main steps for carrying out GA are as follows:

1. **[Start]** Generate random population of *n* chromosomes (suitable solutions for the problem)
2. **[Fitness]** Evaluate the fitness *f(x)* of each chromosome *x* in the population
3. **[New population]** Create a new population by repeating following steps until the new population is complete
    1. **[Selection]** Select two parent chromosomes from a population according to their fitness (the better fitness, the bigger chance to be selected)
    2. **[Crossover]** With a crossover probability cross over the parents to form a new offspring (children). If no crossover was performed, offspring is an exact copy of parents.
    3. **[Mutation]** With a mutation probability mutate new offspring at each locus (position in chromosome).
    4. **[Accepting]** Place new offspring in a new population

4. **[Replace]** Use new generated population for a further run of algorithm
5. **[Test]** If the end condition is satisfied, **stop**, and return the best solution in current population
6. **[Loop]** Go to step **2**

Case Study

Oil refinery consists of different units and sources of air emissions. A typical simple refinery layout was taken as an example and ten different sources were identified and their emissions were reported as in the table below.

Table 2: Sources with air emissions

Source	NO_x (ppm)	SO_2 (ppm)	CO (ppm)
1	128	287	0
2	116	260	0
3	96	260	0
4	45	12	30
5	52	8	22
6	50	2	6
7	45	8	16
8	42	3	3
9	55	5	8
10	40	107	6

GA is applied targeting two objective functions dealing with minimizing the air emissions below the standard and the cost. For this problem, the length of chromosomes is the number of refinery's air emission sources which consists of individual genes. Each gene yields different controllers that can be used for related source. The initial population of the chromosomes was generated randomly. The fitness of each chromosome could be obtained by calculating the weighted value of cost and air emission concentration. The objective function presented in this paper is shown in Equation 3.

$$fitness(p) = W_{cost}\frac{\log_{10}(FC(p))}{\log_{10}(Cost_{\max})} + W_{pollution}\frac{\log_{10}(FP(p))}{\log_{10}(V_{CO}^2 + V_{NO_x}^2 + V_{SO_2}^2)} \qquad (3)$$

p : evaluated chromosome,

$FC(p)$: controller's cost of chromosome p,

$FP(p)$: weighted value of pollutants concentration on the ground

W_{cost} : cost weight

$W_{pollution}$: pollutants weight.

$W_{pollution}$ has more impact value than W_{cost} because air pollution control is more important than cost reduction. The value of these weight functions was settled by calculating average and variance of objective function for each value illustrated in Table 2. So the best cost and air pollutant weight were 0.2 and 0.8 respectively to have minimum value.
The amount of emissions (SO_2, NO_x, and CO) on the ground level after considering proposed controllers by utilizing the model developed was calculated and then *FP* was found based on equation 4. If each of the pollutant's concentration is more than its standard, penalty value is calculated for related pollutant by utilizing equations (a) to (c) below.

$$FP(p) = (V_{CO} + P_{CO}) + (V_{NO_x} + P_{NO_x}) + (V_{SO_2} + P_{SO_2}) \tag{4}$$

Where V_{CO}, V_{SO_2} and V_{NO_x} are the pollutants volume on the ground level and P_{NO_x}, P_{SO_x}, P_{CO} are penalties of each pollutant which were computed by the following equations.

$$P_{CO} = \left(V_{CO} - V_{CO\_Standard}\right)^2 \tag{a}$$

$$P_{SO_2} = \left(V_{SO_2} - V_{SO_2\_Standard}\right)^2 \tag{b}$$

$$P_{NO_x} = \left(V_{NO_x} - V_{NO_x\_Standard}\right)^2 \tag{c}$$

After computing objective function for all existing chromosomes in each generation, the genetic algorithm was utilized and the chromosomes were sorted in order of increasing objective function. The best two chromosomes were transferred to new population. The couple was selected from mating pool by tournament selection and offspring produced by two point crossover technique. Mutation is accomplished by deliberately changing one gene in a random fashion. The whole process was repeated for predefined number of generations.

Results and Discussion

Air emissions control within an oil refinery is considered to be a challenging issue since the aim is to minimize cost and at the same time, minimize air emissions. An optimization approach is carried out in this study using genetic algorithm for a typical simple oil refinery layout. Table 3 summarizes the control option for each source mentioned earlier. The model does not allow applying more than a control technology on each source.

Table 3: Control option for each source

Unit.	Control
1	SNCR
2	Wet Scrubber
3	Wet scrubber
4	Thermal incinerator
5	No control
6	Low-NO_x Burner
7	No control
8	Thermal incinerator
9	No control
10	No control

The fitness function is an important factor for convergence and stability of genetic algorithm. Table 4 shows average and variance of objective function for various weight functions for the cost objective.

W_{cost}	Variance of fitness function	Average of fitness function
0.3	10.2E-4	0.47
0.2	8.5E-4	0.45
0.1	9.9E-4	0.46

The maximum concentration of SO_2, NO_x, and CO on the ground level before and after considering control options is calculated. The concentration of pollutants was reduced over 90 % and all of them are below standard value. Figure 1 shows the results for SO_2 as an example before and after application of control options.

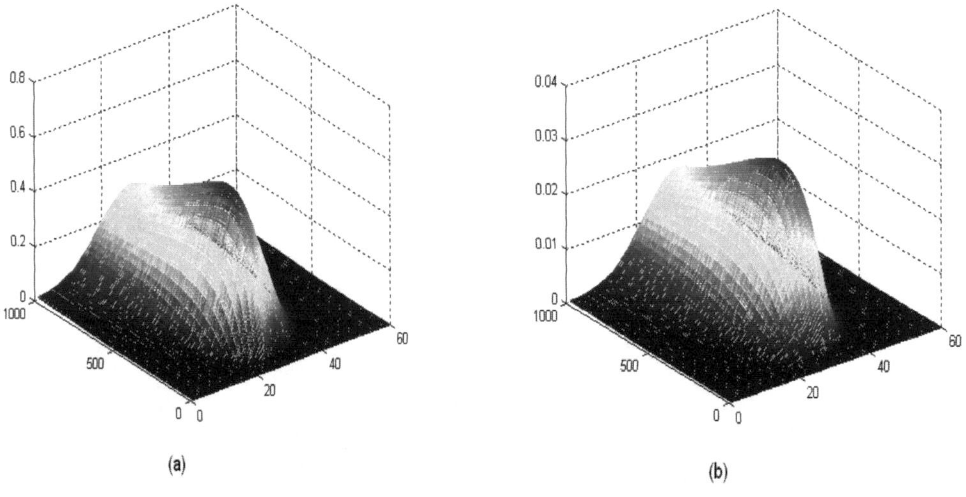

Figure 1: SO_2 emissions before (a) and after (b) considering control options

Conclusion

A genetic algorithm approach was carried out to find the suitable air emission control option for different sources within a simple typical refinery. Three emissions were considered in this study and the results showed that some sources that emit a certain emission in a quantity more than the standard needs to be controlled by applying the appropriate control option in terms of cost and efficiency. A significant decrease in the emissions was noticed when control options were implemented.

Acknowledgment

The author would like to thank KFUPM for their support.

References

[1] R. J. Heinsohn, R. L. Kabel, Sources and Control of Air Pollution. Prentice Hall, New Jersey, 1999.
[2] EPA, Air Pollution Control Cost Manual, sixth Ed, Environmental Protection Agency, EPA 452/B-02-001, Research Triangle Park. NC, USA, 2002.
[3] C.D. Cooper, Air Pollution Control a Design Approach, Waveland Press, Long Grove, Illinois, USA, 2002.
[4] S. Salvador, Y. Kara, J.M. Commandre, Factors influencing pollutant gas emissions of VOC recuperative incinerators—Large-scale parametric study, Applied Thermal Eng. 26 (2006) 1640–1651.
[5] Wikipedia, the free encyclopedia, 2010. Available at: http://en.wikipedia.org/wiki/Air_pollution, this page was last modified on 18 February 2010.
[6] A. Elkamel, E. Fatehifar, M. Taheri, M.S. Al-Rashidi, A. Lohi, A heuristic optimization approach for Air Quality Monitoring Network design with the simultaneous consideration of multiple pollutants, J. of Environmental Management. 88 (2008) 507–516.
[7] G.V. Reklaitis, A. Ravindran, K.M. Ragsdell, Engineering Optimization: Methods and Application. John Willy, New York, 1983.
[8] N.D. Nevers, Air Pollution Control Engineering. McGraw-Hill, New York, 2000.
[9] J.R. Jang, C.T. Sun, E. Hizutani, Neuro-Fuzzy and soft computing. Prentice Hall, New Jersey, 1997.
[10] H. S. Peavy, S.R. Rowe, G. Tchobanoglou, Environmental Engineering. McGraw-Hill, New York, 1985.
[11] R.L. Haupt, S.E. Haupt, Practical Genetic Algorithms. John Wiley & Sons, New York, 1998.
[12] M. Melanie, An Introduction to Genetic Algorithms. Massachusetts Institute of Technology, London, England, 1996.

Pretreatment of high concentration caustic waste wastewater from catalytic gasoline process unit in refinery

Qiu Li-ping★ and Luo De-chun, (College of Environmental Science and Engineering, Chang' an University, Xi' an, Shaanxi, China)

Abstract: The high concentration caustic waste wastewater of constituents complexity、heavy pollution and poor biodegradability of catalytic gasoline process unit in refinery were studied adopting the pretreatment of desulfurization (self-developed desulfurization agent), acidification(to recovery phenol) and the oxidation-coagulation of hydrogen peroxide- ferrous sulfate（H_2O_2-$FeSO_4$, Fenton reagent）. The results of experiment showed that the degree of contamination of caustic waste wastewater was debased greatly by the process. The removal efficiencies of chemical oxygen demand（CODcr）、sulfide、phenol and chromaticity reached 86.88%, 99.54%, 98.10% and 97.14% respectively. The biodegradability was improved significantly and B / C ratio was increased from 0.24 to 0.51 when desulfurization occurred for 48 hour and oxidation-coagulation occurred for 60 minutes at pH=6-7.Therefore, the CODcr loading rate in downstream biological treatment system was reduced greatly for incoming caustic waste wastewater.

Key words: Caustic waste wastewater. Pretreat- ment. Desulfurization agent. Fenton reagent. Oxidation-coagulation process

1 Introduction

The caustic waste wastewater with strong alkaline discharged from catalytic gasoline process unit in refinery generally have many pollutants of hard biochemical degradation and some of them such as $COD_{cr,}$ sulfide and phenol in the stream account for 40%-50% (A Kolhatkar et al. 1995; Wujing 2007) of total amounts of pollutants. In addition, also contains a lot of naphthenic acid and other organic matter (Ye Zong-jun et al. 2006). If those pollutants are treated unreasonably or directly released to wastewater treatment plant (WWTP), they would result in the failure of biochemical treatment systems, reduce the removal efficiencies and increase the operating costs, and cause the adversely impact on the performance of WWTP. Considering the high toxicity of caustic waste wastewater, the WWTP would be vulnerable to the impact of high strength stream and reduced effluent quality if certain wastewater is directed into treatment units. Therefore, the most efficient methods should be considered to pre-treat the caustic waste wastewater prior to be discharged from refinery. At present, caustic waste wastewater treatment methods commonly used are neutralization, oxidation and biological treatment.

Neutralization methods mainly have neutralization of carbon dioxide and Sulfuric acid (Yu Wei 2001; Bai Yao-huai 1999; Li Shuchun et al. 2004). Carbon dioxide method use the waste carbon dioxide, which reflects the optimization of resource use, but the method is large in one-time investment of equipment, and not suitable to treat small batch processing of the catalytic gasoline caustic waste wastewater discharged periodically, and sodium carbonate produced by the technology is lower-purity and bad sales , in particular, there is the serious pollution of the exhaust gas of the spray-dried sodium carbonate in the production process, which lead to these devices often being half lay-off status and affect the treatment of caustic waste wastewater. Sulfuric acid neutralization method acidify generally caustic waste wastewater from 2 to 4 of pH to recovery phenols substances (Song

Yongxin. 2006), which greatly reduce the phenol and COD in wastewater (Ni Binghua et al. 2006), but the Na_2SO_4 produced after acidification easy form crystal and seriously affect subsequent processing, moreover the caustic waste wastewater containing high sulfide will lead to a large number of H_2S gas, which is very adverse for the subsequent processing of sulfides in caustic waste wastewater. Most of the pollution substances can be removed after the pretreatment by acidification of wastewater, but phenol effluent concentrations are still very high, up to thousands of to tens of thousands of mg/L(Zhao Hua et al. 2009)

Oxidation process is divided into air oxidation and chemical oxidation. Caustic waste wastewater oxidized is again pretreated to reduce COD load by sequencing batch activated sludge process (SBR),then wastewater after pretreatment is treated in WWTP (Zhang Xiheng 1996; Liu Yogsong 1990; Ma Fengrong et al. 2006) , which can effectively oxidize the catalytic gasoline waste lye , liquid hydrocarbon waste lye and other high concentrations wastewater. But air oxidation need large energy consumption and chemical oxidation needed high reagents cost, so direct the oxidation treatment of caustic waste wastewater containing a large number of organic compounds obviously is no reasonable in economic.

Biological treatment method is the process that the organic matter and other pollutants in wastewater are degraded and translated into harmless substances by the microbial metabolism (Ding Jiuliang 2001; Li fei 2006; Cai Qingling 2007), and have many advantages in treating caustic waste wastewater in refinery such as low cost , mild conditions, low energy consumption, low investment, no requiring a catalyst, no producing chemical sludge, small footprint, high processing efficiency, operation of stable and safe, etc., so, the method have broad application prospects. But the biological treatment of caustic waste wastewater in refinery still have many problems to be solved, on one hand, conventional biological treatment process is more difficult due to the toxic effects of the high concentrations sulfide and phenol pollutants on microorganisms in the conventional biological treatment system of wastewater. On the other hand, high efficient bacteria screening, microbial community structure and pollutant degradation mechanism etc. still need further studied (Chao Qun-fang et al. 2009). In this study, caustic waste wastewater was first desulfurized by self-developed desulfurization agent, then phenol of wastewater was recovered by acidification, finally, wastewater was pretreated by the oxidation-coagulation of H_2O_2-$FeSO_4$ (Fenton reagent). Based on the above experiments, the desulfurization principles、the effects of acidification recovering phenol and the biodegradability (BOD_5 / CODcr, abbreviated as B / C)of after wastewater treatment were explored, and determined the optimum pretreatment conditions that facilitate effective implementation of the follow-up processing.

2 Experimental

2.1 Materials and instrument

Concentrated sulfuric acid (H_2SO_4, mass percentage 98.3%, analytically pure), high efficient compound desulfurization agent (self-developed), hydrogen peroxide (H_2O_2, mass percentage 30%, analytically pure). Ferrous sulfate heptahydrate ($FeSO_4 7H_2O$, analytically pure, make it to the solution of mass fraction 10% when be used). Polyacrylamide (PAM, analytically pure, average relative molecular mass as1.5×10^7～2.0×10^7 ,make it to the solution of mass fraction 1‰ when be used).

SHIMADZU UV vensus spectrophotometer (UV-2001). ORION pH meter, a set of color-comparison tube(50ml), reflux units for CODcr, a set of distillation devices for volatile phenol, titration table,IR3-6 blender for coagulation, aeration pump, HH-4 isothermal water-bath pot etc..

2.2 Sampling

The caustic waste wastewater (pH>13, chromaticity =2100 times) used in this study were obtained from catalytic gasoline process unit in refinery, in which the major pollutants were CODcr (234,400mg/L),sulfide (12,022mg/l), and phenol(53,202mg/L).

2.3 Test Methods

The precipitate produced by adding concentration of 2-3 g/L desulfurization agent in caustic waste wastewater was separated from wastewater at 40℃ and micro aeration of 48h, from which desulfurization agent recovered could be recycling. The wastewater after desulfurization was immediately stratified by sulfuric acid acidifying to pH=6-7, sitting for a while, the phenol of upper organic phase was recovered in the recovery device, and the under aqueous phase was treated by oxidation–coagulation of Fenton reagent (P K Malik et al. 2003; Deng Xiaohui et al.2007) to investigate the effects of the dosages of H_2O_2 and $FeSO_4.7H_2O$ and pH value and other factors on removal of CODcr, sulfide, phenol, color etc., and to explore the appropriate process conditions and parameters of Fenton oxidation-coagulation method (Kang N et al. 2002).

Analysis methods were presented as follows: CODcr and sulfide were respectively determined by potassium dichromate method and iodimetry. Phenol was measured by using 4-aap and BOD_5 was determined by five-day biochemical cultivation. Chromaticity was measured by the method of dilution multiple.

3 Results and discussions

3.1 The Effect of Desulfurization and Acidification on Caustic Waste Wastewater

3.1.1 The Principles for Desulfurization

High-efficient compound desulfurization agent is a kind of polyelectrolyte coagulants made from multiple high polymers through complex physical and chemical reactions，which is characterized by superior ability of oxidization、coagulation and desulfurization. The sulfide in caustic waste wastewater, existing as the salt form of most NH_4^+ and Na^+, react chemically first with high valence metal ions (M^{x+}, x： represents valence) produced by hydrolysis of desulfurization agent. The reaction equation is presented as follows:

$$2M^{x+}+xS^{2-}\rightarrow M_2S_x\downarrow \qquad ①$$

At the same time，desulfurization agent also produce a variety of high-valence metal hydroxyl ions ($M(OH)^{(x-1)}$ 、 $M(OH)_2^{(x-2)}$ …) by hydrolysis and the causing turbidity substances of oil-soluble activity, which occur coagulation reaction with sulfide in wastewater ,and form polymer chain under the role of adsorption and bridging , which form floccules of large particle by intercross each other, then floccules are precipitated and separated，finally achieve the purpose of removing sulfide.

In addition，a small part of sulfide directly exposed in the air are oxidized into $S_2O_3^-$，then $S_2O_3^-$ are further oxidized into SO_4^{2-}. The reaction equations are presented as follows:

$$2HS^-+2O_2\rightarrow S_2O_3^{2-}+H_2O \qquad ②$$

$$2S^{2-}+2O_2+H_2O\rightarrow S_2O_3^{2-}+2OH^- \qquad ③$$

$$S_2O_3^{2-}+2O_2+2OH^-\rightarrow SO_4^{2-}+H_2O \qquad ④$$

Disulfide (RSSR), produced after thiols organic substances are oxidized, can be separated from wastewater. The reaction equation is presented as follows:

$$2RSNa+2R^1SNa+O_2+2H_2O \rightarrow 2RSSR^1+4NaOH \quad ⑤$$

3.1.2 Recover phenol by acidification

There are a large number of sodium phenolate in catalytic gasoline waste wastewater, when micro-aerated, phenol produced after acidification of wastewater was recovered. The reaction equation is presented as follows:

$$Ar\text{-}ONa + H^+ \rightarrow Ar\text{-}OH + Na^+ \quad ⑥$$

The remaining $S_2O_3^{2-}$ in the wastewater would react as follows:

$$S_2O_3^{2-} + 2H^+ \rightarrow H_2S_2O_3 \quad ⑦$$

H2S2O3→H2SO3+ S↓ ⑧

$$H_2SO_3 \rightarrow SO_3^{2-} + 2H^+ \quad ⑨$$

2SO32-+ O2→2SO42- ⑩

3.1.3 The Effect of Desulfurization and Acidification

Analysis data for desulfurization and acidification of wastewater were presented in table 1:

Table 1 The effect of desulfurization and acidification on removal efficiencies of pollutants

Parameters	pH	CODcr(mg/L)	Sulfide(mg/L)	Phenol(mg/L)	BOD_5(mg/L)	chromaticity (multiples)	B / C
Before desulfurization and acidification	> 13	234400	12022	53202		2100	
After desulfurization and acidification	6-7	65854	338	2172	15805	1050	0.24
Removal (%)		71.91	92.6	83.55		50	

From Table 1, it could been observed that the removal efficiencies for CODcr, sulfide, phonel and chromaticity were satisfactory by From Table 1, it could been observed that the removal efficiencies for CODcr, sulfide, phonel and chromaticity were satisfactory by desulfurization under micro aeration, for the reasons that, on one hand, micro-aeration would moderately agitate wastewater so as to make desulfurization agent more readily mixed with wastewater. On the other hand, plenty of sulfides and some small molecule compounds of high concentration such as ammonia and hydrogen sulfide, etc. decomposed by organic compounds containing heteroatom organics in wastewater (Zhong Huawen et al. 2003) would affect the desulfurization rate, but a small portion of oxygen provided by micro-aeration could accelerate the oxidation and desulfurization of sulfides.

3.1.4 The Biodegradability of Caustic Waste Wastewater

The pollution loads of wastewater after desulfurization and acidification(to recovery phenol) were largely reduced, but biodegradability (BOD/CODcr=0.24) was still low and could not be directly treated by aerobic biochemical. When the high strength wastewater was directly discharged into WWTP, it would affect the normal operation and reduce the treatment efficiencies. Therefore, In order to reduce the pollution degree of wastewater, H_2O_2-$FeSO_4$ (Fenton reagent) was used to oxidize and decompose not easy biochemical degradation substances in wastewater after desulfurization and acidification so as to increase theirs biodegradability and reduce the pollution load for the downstream processes.

3.2 H_2O_2-$FeSO_4$ (Fenton Reagent) Effect

3.2.1 The Effect of H_2O_2 Dosage on Pollutants Removal Efficiencies

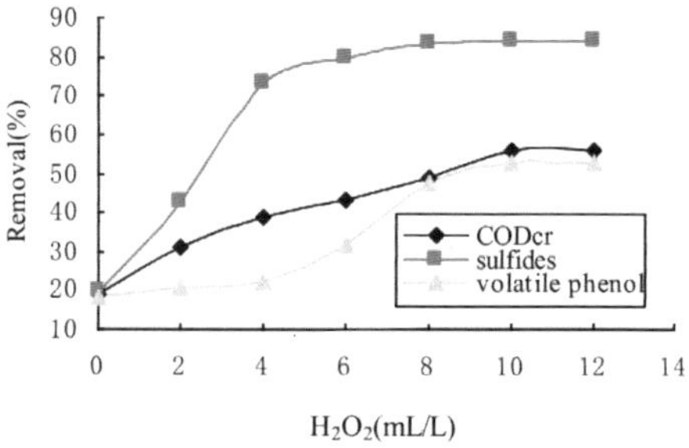

Figure1.The effect of H_2O_2 dosage on the Pollutant Removal Efficiencies

6 parts 200mL samples taken from wastewater after desulfurization and recovering phenol, which contain 5g/L $FeSO_4 \cdot 7H_2O$ and pH 6-7 determined after the preliminary testing and different dosages H_2O_2, were tested respectively in IR3-6 blender for coagulation at 200r/min for 60 min, then were tested again after adding 5mg / L PAM at 100r/min for 1 min, and determined supernatant liquid after precipitation 10 min. The influences of different dosage H_2O_2 on the removal of CODcr, sulfides and volatile phenol were investigated. Experiment Results were presented by Figure 1.

As could be seen from Figure 1, after Fenton oxidation-coagulation process, the wastewater had a good treatment performance, this was the reason that Fenton reagent could oxidize and polymerize water-soluble、colloidal、 difficult degraded and suspended pollutants in wastewater (Zhong Huawen et al. 2003; Li Ya-jun et al. 2009) to form polymerization products that would be easily removed by coagulation. Floccules occurred as a result from agglomeration and flocculation of these polymerization products (Chen Wensong et al. 2004). Experiment results showed that the optimal H_2O_2 dosage was 10mg/L.

3.2.2 The Effect of $FeSO_4 \cdot 7H_2O$ Dosage on Pulltants Removal Efficiencies

According to the above test results, the impacts of different dosages $FeSO_4 \cdot 7H_2O$ on the removal of CODcr, sulfides and volatile phenol were investigated when the concentrations of H_2O_2 was maintained at 10 mg/L (and other procedures and conditions as above). Experiment Results were presented by Figure 2.

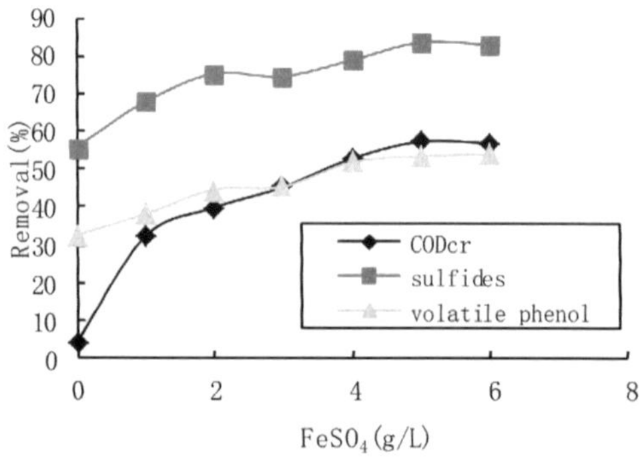

Figure2. The effect of $FeSO_4 \cdot 7H_20$ dosage on the Pollutant Removal Efficiencies

From Figure 2, it showed that the removal efficiencies of CODcr, sulfide and phenol were higher when the concentration of $FeSO_4 \cdot 7H_2O$ was increased to 5g/L. However, when mass concentration was increased to 6g/L, wastewater would become turbidity and removal efficacies would decrease. This was the reason that more $Fe(OH)_3$ colloid would be generated by oxidation and hydrolysis with the increase dosage of $FeSO_4 \cdot 7H_2O$, and some of them and polymerization products formed were removed by co-precipitation, but $Fe(OH)_3$ colloid without causing co-precipitation led to turbidity of the solution. Therefore, the optimal $FeSO_4 \cdot 7H_2O$ dosage was 5g/L.

3.2.3 The Effect of pH Value on Pollutants Removal Efficiencies

Based on the experimental results above, when the concentrations of H_2O_2 and $FeSO_4 \cdot 7H_2O$ were maintained at 10 mg/L and 5 g/L, respectively (other procedures and conditions as above), the effect of different pH value on the removal efficiencies for various pollutants were investigated and shown in Figure 3.

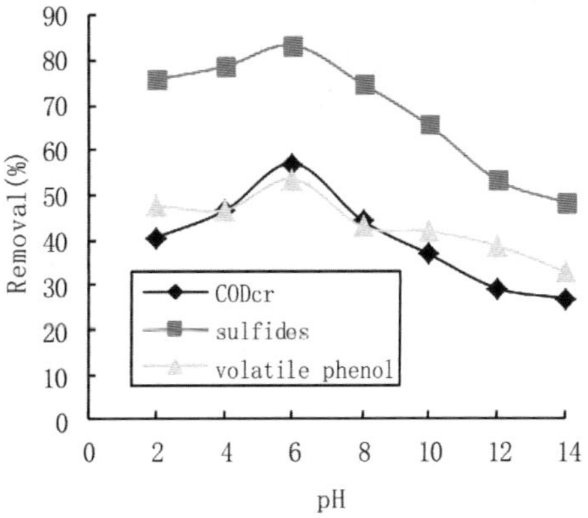

Figure3. The effect of pH values on Pollutant Removal Efficiencies

From Figure 3, it showed that pH values had strong influence on the removal efficacies for pollutants.

When PH≤6, pollutants removal efficiencies were relatively higher. When pH=6, the removal efficiencies reach the highest point even without occurring any floccules precipitations. When pH >6, the removal efficiencies for various pollutants were reduced with pH increasing. The reason was that varying pH values would determine either Fenton oxidation or Fenton coagulation to be dominated (Liu Yongdi et al. 1994). When pH value was lower, a kind of hydroxyl free radical (• OH) of strong oxidation capacity was produced by H_2O_2 under catalytic action of Fe^{2+} in Fenton reagent, which make carbon chain of organic pollutants in wastewater oxidized and cracked, and difficult biodegradable organic macromolecules cracked into readily biodegradable small molecules organic matters, or completely mineralized into CO_2 and H_2O. So, low pH value would cause Fenton oxidation and otherwise resist to Fenton coagulation that would cause floccules formation. The oxidation and coagulation would be occurred at pH=6-7, at this time, organic compounds in the wastewater could be oxidized and coagulated synchronously and form some polymerization products of relative small molecular mass so that the solubility and settledbility of these pollutants would be improved, associated with cohesion of Fe^{2+}, the co-action of Fenton oxidization and coagulation would result in an increasing of the removal efficiencies for CODcr, sulfides, phenol and chromaticity (Zhu Wanpeng et al. 1995). When pH>7, Fenton oxidization stopped and inhibited the oxidization and polymerization of soluble contaminations, so the solubility and settledbility remain the same, at this time, there were floccules, but the major floccules were $FeSO_4$ and PAM themselves. Therefore, the removal efficiencies were lower for all pollutants. Experiment results showed that the optimal pH=6-7.

3.2.4 The Effect of PAM Dosage on Pollutants Removal Efficiencies

when the concentrations of H_2O_2 and $FeSO_4 \cdot 7H_2O$ were maintained at 10 mg/L and 5 g/L, respectively (other procedures and conditions as above), the effect of different dosages PAM on the removal efficiencies for various pollutants were investigated. Experiment results showed that PAM of 5mg / L could result in higher removal rate of pollutants. But continue to increase PAM dosage would cause slightly decreasing of pollutant removal efficiencies. The reason was that the excessive PAM not only would result in toxicity and cause more CODcr, but also increased the cost of processing. Therefore, PAM concentration of 5mg / L was preferred.

3.2.5 The Biodegradability of Pre-treated Wastewater

Based on the study of constituent complexity and poor biodegradability of the caustic waste wastewater from catalytic gasoline process unit in refinery, the results obtained were satisfactory and were presented in Table 2.

Table 2 The performance results of oxidation-coagulation

items	pH	CODcr(mg/L)	Sulfide(mg/L)	Phenol(mg/L)	BOD_5(mg/L)	Chromaticity (multiples)	B / C
After desulfurization and acidification	6-7	65854	338	2172	15805	1050	0.24
Fenton treatment	6-7	30765	54.9	1011	15823	60	0.51
Removal（%）		53.28	83.73	53.45		94.28	
Total removal（%）		86.88	99.54	98.10		97.14	

From Table 2, it could be seen that, after the pretreatment of caustic waste wastewater, BOD/CODcr ratio increase from 0.24 (after desulfurization and acidification) to 0.51(after Fenton oxidization-coagulation).

This was main reason that most of complex large organic molecules were degraded into small molecular pollutants in wastewater after pretreatment by Fenton reagent， therefore， the biodegradability of wastewater was improved significantly. then wastewater that a small amount of the precipitate produced was separated was further processed in biochemical system， which would not impact the follow-up biochemical system and affect the normal operation of the WWTP. (Ma Chengyu et al. 2007).

4 Conclusions

Due to the complexity of constituents, heavy pollution and low biodegradability, the high- efficient desulfurization agent self-developed and the sulfuric acid were used to desulfurized and acidify caustic waste wastewater under the condition of micro-aeration. Results showed that the process was effective and the removal efficiencies of CODcr, sulfide, phenol and chromaticity reached 71.91%, 92.6%, 83.55% and 50%, respectively.

The wastewater after desulfurization and acidification (recycling phenol) were further oxidized and coagulated by Fenton reagent under the condition of pH = 6-7 and adding H_2O_2 of 10mL / L, $FeSO_4 \cdot 7H_2O$ of 5g / L and PAM of 5mg / L respectively, the removal efficiencies for various pollutants would be improved significantly. Not only the removal efficiencies of CODcr, sulfides, phenol and chromaticity in wastewater were again raised 53.28%, 83.73%, 53.45% and 94.28% respectively, but also the biodegradability was improved and the BOD/CODcr from 0.24(after desulfurization and acidification) to 0.51(after Fenton oxidization-coagulation) as well.

Acknowledgments This research was supported by the teachers of Engineering Research Center for Groundwater and Eco-Environment of Shaanxi Province， Here to express my sincere appreciation to them.

References

A Kolhatkar，K Sublette（1996）. Biotreatment of refinery spent sulfidic caustic by specialized cultures and acclimated activated sludge. Applied Biochemistry and Biotechnology58(57)：945-946

Wujing(2007)Pretreatments technologies for high-concentrated caustic sludge wastewater of petroleum chemical industries. Modern Business Trade Industry 19(3): 153-154

YE Zong-jun, XU You-hao, WANG Xie-qing (2006)Study on the composition of products from catalytic cracking and thermal cracking for heavy fractions of FCC gasoline. ACTA PETROLEI SINICA (PETROLEUM PROCESSING SECTION). 22(3):46-53.

Yu wei (2001) Refinery alkaline residue treatment and comprehensive utilization. ENVIRONMENT PROTECTION IN PETROCHEMICAL INDUSTRY 24(2)：43-47

Bai Yuchua, Bai Yunhong,Lei Xunmao(1999) Application of Additive Carbonized Method to Treating Reclaimable Alkali Residue ENVIRONMENT PROTECTION IN PETROCHEMICAL INDUSTRY 22(1)：55-57

Li Shuchun,Guo Xiwen(2004)Gasoline alkaline residue treatment and recycling production test ENVIRONMENTAL PROTECTION OF CHEMICAL INDUSTRY24(S1):316- 317

Song Yongxin.(2006) Refinery solid waste generation and disposal. SCIENCE & TECHNOLOGY INFORMATION4(20):125

Ni Binghua, Huang Fenglin, Liu Jurong(2006) KINETIC RESEARCH OF CAUSTIC REGENERATION IN FCC GASOLINE CAUSTIC REFINING PETROLEUM REFINERY ENGINEERING 36(4):55-59

ZHAO Hua，CAO Qing-qing，LIU Kan(2009) Status and development of treatment technology for alkaline waste from oil refinery. SCIENCE & TECHNOLOGY IN CHEMICAL INDUSTRY．17(2)71-74

Zhang Xiheng(1996)Wastewater anaerobic biological treatment engineering. China environmental science press .Beijing, China.331 pages

Liu Yogsong(1990)．Treatment of phenol—containing waste water by sequencing contact stabilization process[J]．Water Treatment 5(1)：77-86．

Ma Fengrong,Mu Tong,Bai Bing etal (2006)Application of high concentration wastewater treatment technology in diesel oil Caustic sludge treatment. ENVIRONMENTAL PROTECTION OF OIL & GAS FIELDS16(2):19-21

Ding Jiuliang(2001) The application of Biological treatment methods in the Boxi terminal oily wastewater COD treatment process. CHINA OFFSHORE OIL AND GAS(ENGINEERLNG)123(5):1-5

Li fei(2006) The innovation achievement of refining wastewater treated by Microbial method. China Environment News2(7)

Cai Qingling(2007) Status and Prospects of refining caustic sludge processing. LIAONING URBAN AND RURAL ENVIRONMENTAL SCIENCE & TECHNOLOGY27(2):19-24

CHAO Qun-fang,ZHOU Jun,FANG Xin-xiang,etal(2009) Studies on Isolation and Degrading Characteristics of Phenol-degrading Bacteria ZJ-1. BIOTECHNOLOGY19(2):57-59

P K Malik,S K Saha.(2003)Oxidation of direct dyes with hydrogenperoxide using ferrousion as catalyst. Separation and Purification Technology 31(3):241-250.

Deng Xiaohui,Zhang Haitao,Cao Guomin et al (2007)Progress of the research and application of Fenton Reagent for wastewater treatment. Shanghai Chemical Industry 32(8): 1-5

Kang N,Lee D S,Yoon J(2002)Kinetic modeling of Fenton oxidation of phenol and monochlorophenols.Chemo-sphere 47(9)：915-924.

Zhong Huawen, He Dongsheng, Liao Yan(2003)Pollution and pretreatment of refinery basic sewage.Petrochemical Technology & Application.21(4):288-290.

Li Ya-jun, ZHENG Jie(2009)Energy-use improvement of FCC process for crude gasoline upgrading. ACTA PETROLEI SINICA(PETROLEUM PROCESSING SECTION).25(2): 228-232.

Chen Wensong, Wei Chaohai(2004)Study on Fenton oxidation-coagulation process for the treatment of printing and dyeing wastewater.Industrial Water Treatment.24(4):39-41.

Liu Yongdi, Wang Huaxing, Zhu Yaxin(1994)Study on the mechanisms of phenol-bearing wastewater treatment using Fenton oxidization coupling coagulation".China Environmental Science.14(5): 341-345.

Zhu Wanpeng, Yang Zhihua, Wang Li(1995)Application of ferrous-hydrogen peroxide for treatmentof H-acid manufacturing process wastewater. China Environmental Science15(5): 368-372.

Ma Chengyu, Peng Yingli(2007)The treatment and control of high-concentrated refractory organicwastewater. Chemical Industry Press,Beijing,China.1

(135a) Hybrid Processing for CO_2 Capture – A Model Based Approach for Process Synthesis

Anna-Katharina Kunze[1], Philip Lutze[1] and Andrzej Górak[2], (1)Department of Biochemical and Chemical Engineering, Laboratory of Fluid Separations, TU Dortmund University, D-44227 Dortmund, Germany, (2)Department of Biochemical and Chemical Engineering, Laboratory of Fluid Separations, TU Dortmund University, Dortmund, Germany

Hybrid processing for CO_2 capture – A model based approach for process synthesis

A. Kunze, P. Lutze, A. Górak

TU Dortmund University, Laboratory of Fluid Separations

Emil-Figge-Strasse 70, D-44227 Dortmund, Germany

Phone: +49 (0) 231 755-3034, E-Mail: anna-katharina.kunze@bci.tu-dortmund.de

In general, separation of gaseous components is an important process step in chemical processes. But process synthesis for this application is a challenge, because there is a huge variety of possible unit operations to fulfill the same separation task. A restriction to a small number of possible unit operations for gas separations is not that easy, because there is no as dominant unit operation as for example distillation for liquid separations. This leads on the one hand to the question, how to decide which unit operation should be used in each individual separation task and operates best at different process conditions. On the other hand, due to the large number of possible unit operations, hybrid processes might be an opportunity here to gain process intensification. Hybrid processes are defined as combinations of at least two separation techniques in one operation by using the synergy between them [1].

Gas separation processes are often developed based on knowledge [2]. This might end in fixed decision pathways. A model based approach connected to a superstructure analysis will be presented to obtain a higher flexibility and a more general procedure in process synthesis for gas separations. In contrast to purely knowledge based approaches, this method offers the opportunity to take more process configurations into account and evaluate them. A scheme of this multi-step method is shown in Figure 1.

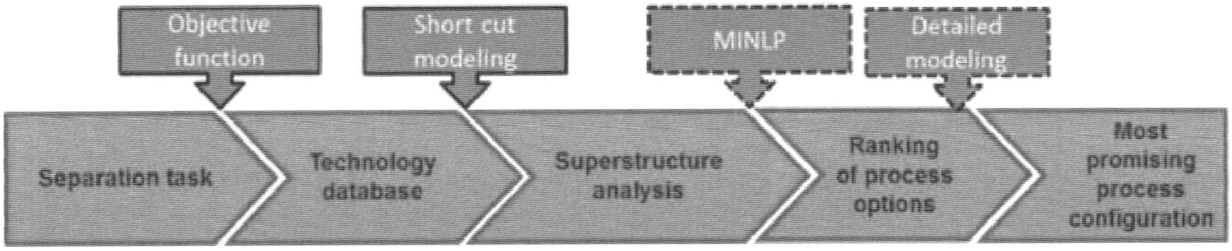

Figure 1: Method scheme for model based process synthesis

Therefore, the first step after defining a separation task is to use a technology database to summarize suitable unit operations. This database contains information about various separation technologies for gaseous mixtures including further reactions and necessary recycle runs. Not only industrial mature technologies can be taken into account, but also promising new technologies like membrane contactors. For each technology the process windows, structural and logical constraints are implemented to restrict the number of process options. The next step is a superstructure analysis based on short cut models of each separation step. If necessary mixed integer non-linear programming (MINLP) has to be used to gain a number of reasonable process options. Those options will be ranked following the objective function, for example energy consumption and the most promising process options are set. In the last step, rigorous modeling can be used to proof the results of the short cut analysis and define the process configurations in a more detailed way.

To figure out a method to find advantageous process configurations, CO_2 capture is used as a case study as a first step. CO_2 can for example be separated with different absorption methods, like washing with aqueous monoethanolamine or piperazine promoted potassium carbonate as well as different absorption technologies such as conventional columns and membrane contactors. But also membrane and adsorption techniques might be suitable to separate CO_2 from flue gas stream. Another possible application of the developed method at a later stage could be the adjustment of biogas upgrading plants to the process conditions of each individual place of action. Therefore, the developed method has to be generally applicable and easily adaptable to other, more complex separation tasks, which include for example more components.

Literature

[1] A. Stankiewicz et al., Re-Engineering the Chemical Processing Plant. Industrial & Engineering Chemistry Research. New York: Marcel Dekker, Inc., 2003

[2] S.D. Barnicki et al., "Separation system synthesis: a knowledge-based approach. 2. Gas/vapor mixtures". Industrial & Engineering Chemistry Research 1992, 31 (7), 1679–1694

(135b) Covalent Organic Polymers (COPs) for Exceptional CO2 Adsorption

Mert Atilhan, Department of Chemical Engineering, Qatar University, Doha, Qatar and **Cafer T. Yavuz**, Graduate School of EEWS, KAIST, Daejeon, South Korea

Efficient CO2 scrubbing without a significant energy penalty remains an outstanding challenge for the fossil fuel-burning industry where aqueous amine solutions are still widely used. One of the main challenges still remains to develop more efficient separation methods that can remove efficiently high fractions of acid gases from natural gas (or any other source) at pre and post combustion streams. For this purpose, for different sources of CO_2 different scenarios are being investigated in order to deal with the removal of CO_2 from various gas streams at various different pressure and temperature conditions. Thus, effective CO_2 capture is essential to investigate the further options to re-use the captured CO_2. Porous materials have long been evaluated for next generation CO_2 adsorbents. Porous polymers, robust and inexpensive, show promise as feasible materials for the capture of CO_2 from warm exhaust fumes. We report the syntheses of porous covalent organic polymers (COPs) with CO_2 adsorption capacities of up to 5616 mg/g (a world record - measured at high pressures, i.e. 200 bar) and industrially relevant temperatures (as warm as 65 ^oC). COPs are stable in boiling water for at least one week and near infinite CO_2/H_2 selectivity is observed. Theoretical calculations refer to an amorphous extended framework as density is likely the main reason for exceptional CO_2 capacities. COPs 1-2 feature basic nitrogen sites that show chemospecific affinity towards acidic gases such as CO_2. COP-3 has reasonably high surface area (418 m^2/g), effective for low pressure operations. Post combustion carbon capture from fossil fuel power plants demands pressures of up to 6 bar and a minimum temperature of 40 ^oC. By tuning their architecture, we show that COPs reach to 3 mmol CO_2/g sorbent at 6 bar and 45 ^oC. High and low pressure capacities make these porous polymer structures viable alternatives to amine scrubbers.

(135c) The Syntheses of CO2 Adsorbent Microporous Materials in Amine Solutions

Fahd Mohamed, Mohammed Shurair, Baligh Dakdakay and Mert Atilhan, Department of Chemical Engineering, Qatar University, Doha, Qatar

This work focuses on the design, production, characterization and testing of materials that potentially have a high capability of capturing carbon dioxide from natural gas precombustion (high pressure adsorption) feed and post combustion (low pressure adsorption) effluent gases. It is anticipated that the proposed work will result in high surface area solid materials that can be utilized for the optimized CO2 capture. Aqueous amine solutions have long been used to remove CO2 and H2S (natural gas sweetening) from industrial gas streams. Alkanolamines such as monoethanolamine (MEA) and N-methyldiethanolamine (MDEA) are some of solvents commonly used for CO2 removal. On the other hand Prussian Blue (PB) analogues have been investigated widely in gas uptake field for their promising gas uptake behaviors. It is known that PB complexes are prepared mostly in aqueous solutions. In this project we will prepare PB complexes in alkanolamine solutions to incorporate these amine molecules into the pores of the aforementioned microporous materials. The amine molecules that occupy the micropores will increase the CO2 adsorption of these materials due to the well-known interaction between CO2 and -NH2 groups. Convenient and conventional amine molecules such as MEA, DEA, and MDEA will be used in this project. Obtained complexes will be characterized with powder X-ray Diffractometer (XRD), Infrared Spectroscopy (FTIR), Thermogravimetric Analysis (TGA) and BET isotherms as full characterization scheme. CO2 adsorption of these clusters will also be investigated using Rubotherm® magnetic suspension balance.

(135d) Development of Stable Multiphase Ni-Mgx-Aly-O Catalysts for CO2 Reforming of Methane

Jun Zhang Sr., Low Carbon Energy Conversion Center, Shanghai Advanced Research Institute, Shanghai, China

Development of stable multiphase Ni-Mg_xAl_yO catalysts for CO_2 reforming of methane

J. Zhang[a], Q. J. Chen[a], B. R. Pan[a], W. B. Kong [a], Q. W. Jin[a], T.J. Zhao*[a], Y. H. Sun*[a,b]

a: Low Carbon Energy Conversion Center, Shanghai Advanced Research Institute, Chinese Academy of Sciences, Shanghai 201203, China

b: State Key Laboratory of Coal Conversion, Institute of Coal Chemistry, Chinese Academy of Sciences, Taiyuan 030001, Shanxi, China

Abstraction:

CO_2 reforming of methane to produce synthesis gas is becoming more research attractive due to the increasing interest in reduction of carbon dioxide emissions and more efficient utilization of natural gas. From an industrial standpoint, Ni-based catalysts have drawn significant research attention for their high activity and wide availability. However, metal sintering, oxidation, and especially significant carbon formation result in the severe deactivation of the Ni catalysts. Some catalysts such as NiO/MgO solid solution, Ni/Al_2O_3, $NiAl_2O_4$ spinel, etc. were reported to have excellent performance for CO_2 reforming of methane. However, those kinds of catalysts still have their own disadvantages, for example, hydrophobia of NiO/MgO, low active site dispersion for Ni/ Al_2O_3, difficult reduction for $NiAl_2O_4$ spinel, etc.

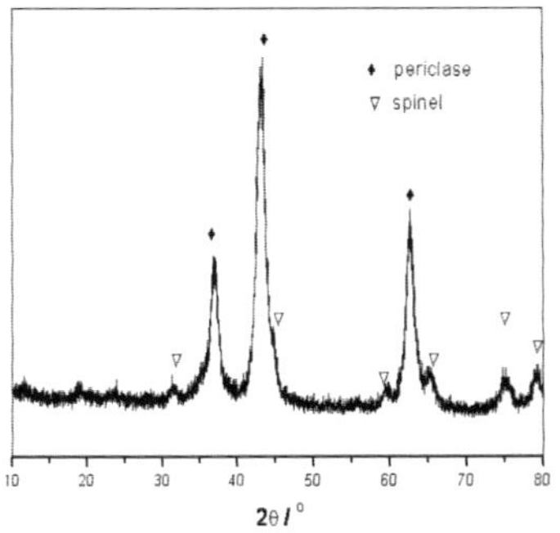

Fig .1 XRD for Ni-Mg_xAl_yO catalysts

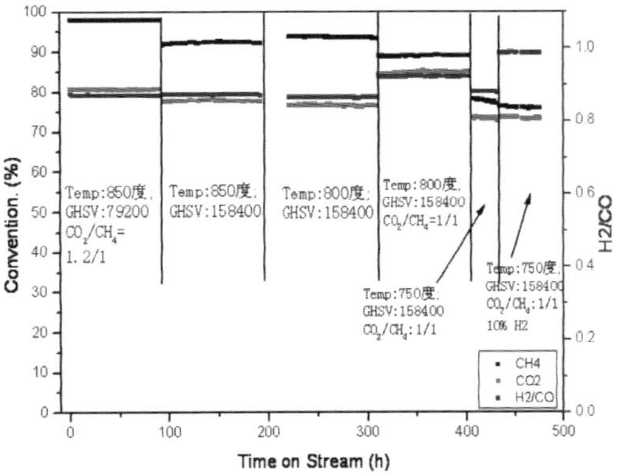

Fig .2 Stability of Ni-Mg_xAl_yO for CO_2 reforming of methane

In this work, a high stable multiphase Ni-Mg_xAl_yO catalyst for CO_2 reforming of methane was developed by HT precursors. For comparison, NiO/MgO solid solution, Ni/Al_2O_3, and $NiAl_2O_4$ spinel catalysts were also prepared and tested. The Catalysts were characterized using ICP-MS, BET, XRD, H2-TPR, TEM, and TG. Catalysts screening for the carbon dioxide reforming of methane showed that Ni-Mg_xAl_yO catalysts with multiphase (periclase-spinel) obtained by hydrotalcite-type (HT-type) precursors exhibited superior performance to other Ni–Mg–Al–O catalysts with different phases. This special multiphase catalyst exhibits very high and stable activity over 1000 h without deactivation at a GHSV of 79, 000 ml/(hg) at 850 °C. Both commercial Ni/Al_2O_3 catalyst and the NiO/MgO solid solution with single periclase phase deactivated quickly after the ignition of reaction, and the $NiAl_2O_4$ spinel show less activity for their low reductively. BET, XRD and H2-TPR results indicated that the activity of the catalysts were strongly depended on the structures, the sizes of the nanoparticles, surface area and active site dispersion. In fact, those characteristics are strongly depended on the types of the phases in the catalyst, and the interaction between them. It was believed that the synergy between periclase and spinel could also significantly improve catalyst performance and reduce carbon formation. A high metal dispersion or small ensemble size can be obtained by NiO/MgO solid solution. The excellent stability of metal was enhanced by spinel phase exist in the NiO/MgO solid solution, which further promoted the metal dispersion and resistance of metal sintering. The synergistic effects between crystal phases and phase cooperation, the phenomena responsible for the variation of activity in the methane reforming as a function of the degree of spinel substitution in periclase were deeply discussed.

Keywords : Natural gas; CO_2 Methane reforming; Ni-Mg_xAl_yO multiphase catalysts; phase cooperation

(137a) Catalytic Conversion Wood Syngas to 'Wide Cut' Diesel Over A Multifunctional Catalyst

Qiangu Yan, Department of Agricultural & Biological Engineering, Mississippi State University, Mississippi State University, MS and Fei Yu, Department of Agricultural & Biological Engineering, Mississippi State University, Mississippi State, MS

A continuous process involving gasification, syngas cleaning, and Fischer–Tropsch (FT) synthesis was developed to efficiently produce 'wide cut' diesel. Oak-tree wood chips were first gasified to syngas over a commercial pilot plant downdraft gasifier. The raw wood syngas contains about 47% N_2, 21% CO, 18% H_2, 12% CO_2, 2% CH_4 and trace amounts of impurities. A purification reaction system was designed to remove the impurities in the syngas such as moisture, oxygen, sulfur, ammonia, and tar. The purified syngas meets the requirements for catalytic conversion to liquid fuels. A multi-functional catalyst was developed and tested for the catalytic conversion of wood syngas to diesel. It was demonstrated that liquid fuels similar to commercial diesel was successfully synthesized from bio-syngas.

MODELO DE CÁLCULO DE LA PERMITIVIDAD DE CARBONES

Peña Benavides, Manuel G. y Vidal Teherán, Dairo A.
Universidad del Atlántico, Barranquilla, Colombia.

Resumen
El carbón posee muchas propiedades que, según su estado, lo hace aplicable para un uso específico. Una de esas propiedades es la permitividad eléctrica, que es la capacidad que posee un material de polarizarse. Esta propiedad es la que fija el comportamiento que van a tener los carbones cuando sean expuestos a una señal electromagnética, marca como se propaga la energía en el material, como se transforma la energía de microondas en calor al interior del mismo y, finalmente, nos indica si un material es susceptible de ser tratado por la vía de microondas. El método de medición de la potencia absorbida, es el que se propone en el presente estudio; este consiste en aplicar al material de análisis una onda electromagnética de una cierta potencia conocida como potencia suministrada para medir la cantidad de esa potencia que absorbe y posteriormente, a través de un modelo matemático, calcular el valor de la permitividad eléctrica.

Palabras claves: *Propiedad dieléctrica, permitividad, microondas, guía de ondas, carbón.*

1. Introducción

Permitividad

La capacidad que posee un material para calentarse en presencia de la acción de las microondas está en función de la permitividad. La unidad de medida en el Sistema Internacional SI, es el faradio por metro (F/m).

Tomada en función de la frecuencia, puede tomar valores tanto reales como complejos, ecuación (1). Por lo general no es constante, debido a que varía con la posición del medio, la frecuencia del campo que se aplique, la humedad del material y la temperatura.

$$\varepsilon = \varepsilon' - i\varepsilon'' \qquad (1)$$

De donde ε' es la constante dieléctrica o permitividad real y, ε'' es el factor de pérdidas dieléctricas o permitividad imaginaria.

Un factor de pérdidas alto indica que el material es susceptible de ser tratado por la vía de microondas.

En un medio que no sea lineal, la permitividad depende de la magnitud del campo eléctrico. En electromagnetismo al campo de desplazamiento eléctrico se le define como D, que representa como un campo eléctrico E influye en la organización de las cargas eléctricas del medio. La relación de estos campos con la permitividad se define con la siguiente expresión:

$$D = \varepsilon \cdot E \qquad (2)$$

En donde el campo de desplazamiento D se mide en culombios por metro cuadrado (C/m^2), mientras que el campo eléctrico E se mide en voltios por metro (V/m) y la permitividad eléctrica ε en faradio por metro (F/m).
En el caso en que ese campo eléctrico se dé en el vacío, la ecuación queda de la siguiente forma [7]:

$$D = \varepsilon_0 \cdot E \qquad (3)$$

De donde, ε_0 es la permitividad del vacío. Dicha permitividad está relacionada con la permeabilidad en el vacío μ_o y la velocidad de la luz C, por la siguiente expresión:

$$\varepsilon_o C^2 \mu_o = 1 \qquad (4)$$

Despejando la permitividad en el vacío,

$$\varepsilon_o = \frac{1}{C^2 \mu_o} \qquad (5)$$

De donde C es la velocidad de la luz ($3*10^8$ m/s) y μ_o es la permeabilidad magnética del vacío ($4\pi*10^{-7}$ Tm/A), definidas en unidades del SI. Se obtiene un valor de la permitividad del vacío de aproximadamente

$$\varepsilon_o = 8{,}8541878176 * 10^{-12}\, F/m$$

Efecto de las microondas sobre la materia

La interacción de las partículas cargadas que poseen ciertos materiales con el campo eléctrico que es generado a causa de la radiación electromagnética que provocan las microondas, hace que estos materiales se calienten.

La irradiación de microondas interactúa con la materia como una fuente interna de energía, haciendo que se caliente directamente desde la profundidad. La eficacia de esta técnica de calentamiento, depende, en gran medida, de las propiedades dieléctricas de la materia.

En el caso de moléculas polares, el campo eléctrico de las microondas hace que los dipolos permanentes presentes en dichas moléculas y los que son inducidos roten al intentar alinearse con el campo oscilante [9]. Para el caso en estudio en el presente trabajo, como se trabaja a una frecuencia de 2,45 GHz, la molécula rota 2450 millones de veces por segundo, y se encontró que a esta frecuencia es ideal para trabajar, debido a que la molécula de agua resuena aproximadamente en ese valor [28].

El carbón y las microondas

La mayoría de los materiales de carbono son excelentes absorbentes de microondas. Las microondas inducen una corriente en fase con el campo electromagnético del material. Los electrones encuentran resistencia al libre desplazamiento por el material, haciendo que cierta cantidad de la energía se disipe en forma de calor.

En los carbones puede ocurrir un interesante fenómeno adicional, la energía cinética de algunos electrones puede llegar a ser suficiente para que salten fuera del material provocando una ionización de la atmósfera que lo rodea [9]. Este fenómeno se puede apreciar con la aparición de chispas o arcos eléctricos.

Los carbones poseen pérdidas eléctricas elevadas y por ese motivo pueden ser sometidos a calentamiento por microondas. Además se pueden utilizar para recibir las microondas y calentar indirectamente otros materiales que no son buenos receptores de microondas, utilizándose en algunos procesos como la pirolisis de biomasa y residuos orgánicos o en reacciones catalíticas heterogéneas.

Guía de onda rectangular

Para la realización del calentamiento de los carbones se hizo uso de una guía de onda rectangular, diseñada y construida especialmente para este estudio.

Tienen varias aplicaciones como en sistemas de radiofrecuencia, microondas terrestres y satelitales. Son fáciles de fabricar y presentan varias ventajas puesto que poseen un gran ancho de banda y presentan pocas pérdidas.

Las corrientes inducidas en sus paredes generan un aumento en las pérdidas de la potencia suministrada, y para contrarrestar esto, la resistencia de las paredes de la guía se debe hacer lo más pequeño posible.

Las paredes deben ser especialmente laminadas, debido a que las corrientes tienden a concentrarse cerca del interior de la superficie de dichas paredes. Estas deben construirse con un material que sea un conductor perfecto.

2. Modelo Matemático para el Cálculo de la Permitividad en Carbones

En el modelo que se plantea con este estudio se realiza un calentamiento por vía de microondas al carbón y se compara con el agua calentada de la misma manera. Se registran los datos de temperatura y tiempo que se grafican y se aplican en el modelo, para el cálculo de la permitividad de los carbones.

Toma como punto de partida el balance de energía del sistema que, en general, está dado por la siguiente ecuación:

$$\{Flujo\ de\ calor\ por\ el\ ingreso\ de\ masa\} - \{Flujo\ de\ calor\ por\ salida\ de\ masa\} + \left\{\begin{matrix} Generación\ de\ energía\ térmica\ por \\ fricción\ molecular \end{matrix}\right\} - \{Calor\ por\ conducción\} = \{Acumulación\ de\ energía\}$$

Reemplazando los términos respectivo

$$F_0 C_P T_0 - F C_P T + 2\pi f V \varepsilon_0 E^2 \varepsilon_c^{''} + 2\pi f V_c \varepsilon_0 E^2 \varepsilon_v^{''} - UA(T - T_a) = M C_P \frac{dT}{dt} \quad (6)$$

Pero como no hay flujo de masa que ingrese ni que salga del sistema, en el caso en estudio, porque se da un calentamiento tipo batch:

$$F_0 = F = 0 \quad (7)$$

El calor generado por la fricción molecular del vidrio, provocado por la acción de las microondas, se puede despreciar debido a que es insignificante con respecto al generado por el carbón. La permitividad eléctrica relativa del vidrio de borosilicato 3.3 que es el tipo del vidrio Duran, es de 0,01702 [38].

El término de acumulación de energía térmica, en función del carbón y del vidrio queda expresado de la siguiente manera:

$$(M_v C_{Pv} + M_c C_{Pc}) \frac{dT}{dt} \quad (8)$$

Por todo lo anterior la ecuación del modelo matemático resulta de la siguiente forma:

$$(M_v C_{Pv} + M_c C_{Pc}) \frac{dT}{dt} = 2\pi f V_c \varepsilon_0 E^2 \varepsilon_c^{''} - UA(T_C - T_a) \quad (9)$$

De donde,
M_v= Masa del vidrio del tubo de ensayo que contiene la muestra del carbón y que se encuentra expuesto a la acción de las microondas, [Kg]
C_{Pv}= Capacidad calorífica del vidrio, [J/Kg °C]
M_c= Masa del carbón, [Kg]
C_{Pc}= Capacidad calorífica del carbón, [J/Kg °C]
$\frac{dT}{dt}$= La variación de la temperatura con respecto al tiempo, [°C/s]
f= Frecuencia de las microondas, [Hz]
V_c= Volumen de carbón, [m^3]
ε_0= Permitividad en el vacío, [F/m]
E^2= Campo eléctrico, [V/m]
$\varepsilon_c^{''}$= Permitividad relativa del carbón
UA= Coeficiente externo de transferencia de calor, [J/s °C]
T_C= Temperatura del carbón, [°C]
T_a= Temperatura ambiente, [°C]

Para el agua se realiza el mismo análisis que se hizo para el carbón obteniéndose la siguiente expresión:

$$(M_v C_{Pv} + M_{H_2O} C_{PH_2O}) \frac{dT}{dt} = 2\pi f V_{H_2O} \varepsilon_0 E^2 \varepsilon_{H_2O}^{''} - UA\left(T_{H_2O} - T_a\right) \quad (10)$$

De donde,
M_{H2O}= Masa del agua, [Kg]
C_{PH2O}= Capacidad calorífica del agua, [J/Kg °C]
V_{H2O}= Volumen de agua, [m^3]
$\varepsilon_{H2O}^{''}$= Permitividad relativa del agua
T_{H_2O}= Temperatura del agua, [°C]

La anterior ecuación posee dos condiciones en sus extremos:

1) Cuando t = 0, entonces, $T_{H_2O} = T_a$

Por consiguiente el segundo término de la derecha de la ecuación (10) se hace cero (0) y se obtiene la siguiente expresión para el t = 0:

$$(M_v C_{Pv} + M_{H_2O} C_{PH_2O}) \frac{dT}{dt}\Big|_{t=0} = 2\pi f V_{H_2O} \varepsilon_0 E^2 \varepsilon''_{H_2O} \quad (11)$$

2) Cuando t = t$_{equilibrio}$, entonces, $\frac{dT}{dt} = 0$

Por consiguiente el término de la izquierda en la ecuación (10) se hace cero (0) y se obtiene la siguiente expresión para el t = t$_{equilibrio}$:

$$2\pi f V_{H_2O} \varepsilon_0 E^2 \varepsilon''_{H_2O} = UA(T_{H_2O} - T_a) \quad (12)$$

De la ecuación (11) se despeja el campo eléctrico y se calcula, ya que los demás términos de la ecuación se pueden determinar o ya se conocen.

$$E^2 = \frac{(M_v C_{Pv} + M_{H_2O} C_{PH_2O})}{2\pi f V_{H_2O} \varepsilon_0 \varepsilon''_{H_2O}} \frac{dT}{dt} \quad (13)$$

De la ecuación (12) se despeja la temperatura de equilibrio del agua y reemplazando el valor del campo eléctrico calculado con la ecuación (13) y haciendo las respectivas operaciones se obtiene el valor de dicha temperatura.

$$T_{H_2O} = \frac{2\pi f V_{H_2O} \varepsilon_0 E^2 \varepsilon''_{H_2O}}{UA} + T_a \quad (14)$$

Esta temperatura del agua debe ser más alta que la temperatura de equilibrio del carbón, debido a que las propiedades dieléctricas del agua hacen que la acción de las microondas sea más fuerte en ella que en el carbón, por esta misma razón se espera que el valor de la permitividad del carbón sea menor que la del agua.

Cálculo de la permitividad del carbón

Hasta el momento se han determinado las ecuaciones que explican el sistema, pero aún falta determinar el valor de la permitividad del carbón.

Para el carbón se dan las mismas condiciones que para el agua, ecuaciones (11) y (12), que para el carbón quedan expresadas de la siguiente manera:

En el arranque $(T_c - T_a) = 0$

$$(M_v C_{Pv} + M_c C_{Pc}) \frac{dT}{dt}\Big|_{t=0} = 2\pi f V_c \varepsilon_0 E^2 \varepsilon''_c \quad (15)$$

En el equilibrio $\left(\frac{dT}{dt} = 0\right)$

$$2\pi f V_c \varepsilon_0 E^2 \varepsilon''_c = UA(T_c - T_a) \quad (16)$$

Despejando de la ecuación (15) la *permitividad en el arranque* del carbón, resulta la siguiente expresión:

$$\varepsilon''_c\Big|_{t=0} = \frac{(M_v C_{Pv} + M_c C_{Pc}) \frac{dT}{dt}\Big|_{t=0}}{2\pi f V_c \varepsilon_0 E^2} \quad (17)$$

De donde,

$\varepsilon''_c\Big|_{t=0} = permitividad\ relativa\ en\ el\ arranque$

Despejando de la ecuación (16) la *permitividad en la temperatura de equilibrio* del carbón, resulta la siguiente expresión:

$$\varepsilon''_c\Big|_{e.e} = \frac{UA(T_c - T_a)}{2\pi f V_c \varepsilon_0 E^2} \quad (18)$$

De donde,

$\varepsilon''_c\Big|_{e.e} = permitividad\ relativa\ en\ la\ temperatura\ de\ equilibrio$

Para poder resolver la ecuación anterior es necesario hacer una condición, que el campo eléctrico se mantiene constante para el carbón y para el agua, ya que la potencia suministrada es mínima, tan solo el 10% de la potencia total del equipo, así mismo, la potencia absorbida por el carbón es pequeña.

3. Gráficas Obtenidas

Se realizaron tres pruebas al agua y a cada uno de los carbones estudiados, bituminoso y antracítico.

Para el Agua

Gráfica 1. Agua Desionizada.

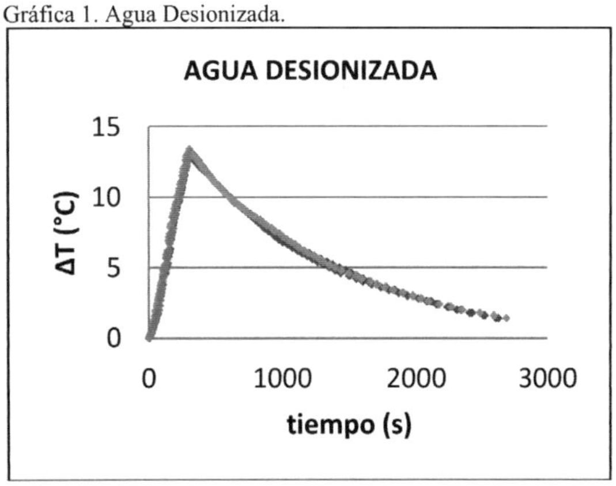

En cada una de las pruebas el calentamiento se realizó durante 5 minutos y se dejó enfriar hasta que el agua alcanzara su temperatura inicial. Se puede observar como el calentamiento se da en forma uniforme para las tres pruebas y que el enfriamiento se da de manera exponencial guardando el mismo perfil.

Para el Carbón Bituminoso

Gráfica 2. Carbón Bituminoso.

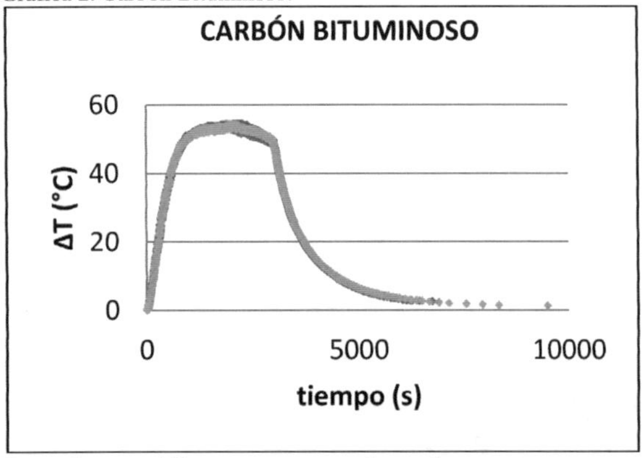

En la gráfica se puede observar como el calentamiento se mantiene constante y después de un tiempo se mantiene a la temperatura de equilibrio, a una diferencia de temperatura entre el carbón y el ambiente se encuentra alrededor de 53°C, y el carbón a 81°C, aproximadamente.

Para el Carbón Antracítico

Gráfica 3. Antracita.

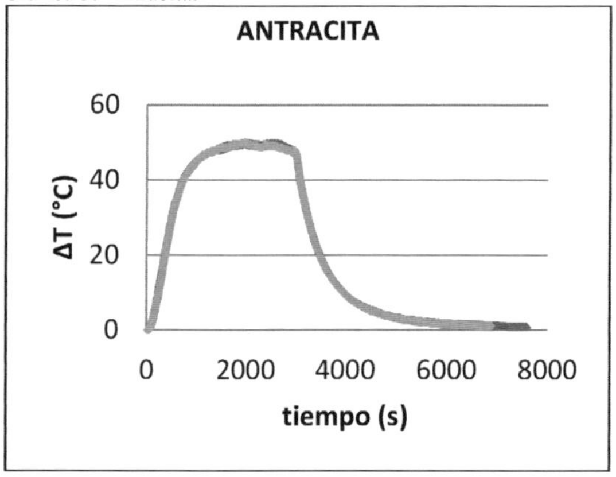

Se puede observar que el calentamiento es de pendiente constante hasta que se comienza a llegar a la temperatura de equilibrio, en donde se mantiene por unos minutos. Pasado un tiempo, el carbón empieza a sufrir daños, haciendo que la temperatura descienda aún con el calentamiento de las microondas.

El enfriamiento se da de manera casi exacta para las tres pruebas, conservando el mismo perfil con buena precisión.

4. Resultados

Luego de que se realizaron las pruebas, se elaboraron las gráficas, se analizaron, se extrajo información de ellas y se aplicó el modelo matemático se obtuvieron los siguientes resultados:

Carbón Bituminoso. Prueba 1.

Campo eléctrico, ecuación (13)

$$E = 370{,}5 \, \frac{V}{m}$$

Temperatura de equilibrio del agua, ecuación (14)

$$T_{H_2O} = 90{,}21740144\ °C$$

Permitividad del carbón en el arranque, ecuación (17)

$$\varepsilon_c''|_{t=0} = 7{,}09196$$

Permitividad del carbón en el estado estacionario, ecuación (18)

$$\varepsilon_c''|_{e.e} = 2{,}19229$$

Carbón Bituminoso. Prueba 2.

Campo eléctrico, ecuación (13)

$$E = 379{,}8\ \frac{V}{m}$$

Temperatura de equilibrio del agua, ecuación (14)

$$T_{H_2O} = 94{,}74207057\ °C$$

Permitividad del carbón en el arranque, ecuación (17)

$$\varepsilon_c''|_{t=0} = 7{,}02674$$

Permitividad del carbón en el estado estacionario, ecuación (18)

$$\varepsilon_c''|_{e.e} = 2{,}10455$$

Carbón Bituminoso. Prueba 3.

Campo eléctrico, ecuación (13)

$$E = 376{,}6\ \frac{V}{m}$$

Temperatura de equilibrio del agua, ecuación (14)

$$T_{H_2O} = 92{,}2738915\ °C$$

Permitividad del carbón en el arranque, ecuación (17)

$$\varepsilon_c''|_{t=0} = 6{,}91878$$

Permitividad del carbón en el estado estacionario, ecuación (18)

$$\varepsilon_c''|_{e.e} = 2{,}02441$$

Antracita. Prueba 1.

Campo eléctrico, ecuación (13)

$$E = 370{,}5\ \frac{V}{m}$$

Temperatura de equilibrio del agua, ecuación (14)

$$T_{H_2O} = 90{,}21740144\ °C$$

Permitividad del carbón en el arranque, ecuación (17)

$$\varepsilon_c''|_{t=0} = 4{,}83344$$

Permitividad del carbón en el estado estacionario, ecuación (18)

$$\varepsilon_c''|_{e.e} = 2{,}54616$$

Antracita. Prueba 2.

Campo eléctrico, ecuación (13)

$$E = 379{,}8\ \frac{V}{m}$$

Temperatura de equilibrio del agua, ecuación (14)

$$T_{H_2O} = 94{,}74207057\ °C$$

Permitividad del carbón en el arranque, ecuación (17)

$$\varepsilon_c''|_{t=0} = 4{,}69888$$

Permitividad del carbón en el estado estacionario, ecuación (18)

$$\varepsilon_c''|_{e.e} = 2{,}41124$$

Antracita. Prueba 3.

Campo eléctrico, ecuación (13)

$$E = 376{,}6\,\frac{V}{m}$$

Temperatura de equilibrio del agua, ecuación (14)

$$T_{H_2O} = 92{,}2738915\,°C$$

Permitividad del carbón en el arranque, ecuación (17)

$$\varepsilon_c''|_{t=0} = 4{,}74887$$

Permitividad del carbón en el estado estacionario, ecuación (18)

$$\varepsilon_c''|_{e.e} = 2{,}51280$$

5. Conclusiones

Fue necesario plantear un modelo matemático, que permitiera plasmar la influencia de la energía transmitida por las microondas a la muestra en la temperatura de esta en el tiempo para poder observar el comportamiento de las mismas, y realizar una deducción de las permitividades de los carbones.

Al comparar los resultados obtenidos con el equipo y modelo matemático planteado, con los obtenidos con anteriormente para el agua, se puede afirmar la validez de los resultados obtenidos en el presente trabajo, ya que se puede apreciar la consistencia de la relación de las permitividades en el estado estacionario e inicial y se mantiene la tendencia a disminuir la permitividad, con un aumento de temperatura a la frecuencia trabajada.

Aunque los resultados obtenidos por medio del equipo construido y el modelamiento matemático planteado, pueden ser analizados y comparados con una tendencia parecida a la del agua, el margen de error que pueda existir está delimitado por la ausencia de un cálculo detallado –con modelos no ideales en el comportamiento del agua en el aumento de temperatura pero esto implica mayor complejidad y tiempo de cálculo.

6. Referencias

[1] ABDELGHANI-IDRISSI, M. A. Experimental investigations of occupied volume effect on the microwave heating and drying kinetics of cement powder in a mono-mode cavity. 2000.

[2] AGUILAR, J. A., GARCIA, R., VALDEZ, Z. Simulación del calentamiento de manganitas mediante microondas. 2007.

[3] AJIACO, F. Evaluación del comportamiento térmico de carbones del Cerrejón, carbones coquizantes y sus mezclas en la producción de coque metalúrgico. Universidad Nacional de Colombia. Bogotá, 2011.

[4] AKHTAR, M. J., FEHER, L., THUMM, M. A Multi-Layered Waveguide Technique for Determining Permittivity and Conductivity of Composite Materials. 2005.

[5] ALMANSA MORALES, Alejandro. Estudio experimental para la optimización energética y termográfica en aplicaciones de calentamiento por microondas. Universidad Politécnica de Cartagena. 2007, 19.

[6] ARIZTIMUÑOZ, A. GONZALEZ, F. RISUEÑO, A. El carbón como materia prima. Escuela Universitaria de Ingeniería Técnica Industrial.

[7] ASTIGARRAGA, Julio. Hornos de alta

frecuencia y microondas. Mc Graw Hill. Bogotá, 1999.
[8] BAKER-JARVIS, J., JANEZIC, M., GROSVENOR, J., GEYER, R. Transmission/Reflection and Short-Circuit Line Methods for Measuring Permittivity and Permeability. National Institute of Standards and Technology. Boulder, Colorado. 1993.
[9] BEVERLY MICROWAVE DIVISION. Magnetron theory of operation. Massachusetts, Estados Unidos.
[10] CARMONA MARTINEZ, F. Análisis y optimización de circuitos de microondas en guía de onda. Universidad Autónoma de Madrid. Madrid. 2009.
[11] CONSTANTINI, S. Guías de ondas. Universidad Metropolitana de Telecomunicaciones.
[12] DILDINE, Bob. Measuring complex permittivity and permeability at RF and microwave frequencies. Hewlett Packard. Yokogawa, Japón. 1989.
[13] FERNÁNDEZ, Y. *et al.* Procesos Térmicos Asistidos por Microondas cobre Materiales Carbonosos. Grupo de Materiales Carbonosos para Aplicaciones Tecnológicas. Oviedo, España. 2010.
[14] GRUPO ESPAÑOL DEL CARBÓN. Boletín del Grupo Español del Carbón. N°16, Mayo 2010.
[15] GUZMAN RABANILLO, D. Sistema de de deposición química en fase vapor para crecimiento homoepitaxial de diamante. Universidad de Cádiz. 2004.
[16]http://translate.google.com.co/translate?hl=es&langpair=en%7Ces&u=http://www.radartutorial.eu/08.transmitters/tx08.en.html
[17] http://www.catamutun.com/produc/carbon/
[18]http://www.lacomunidadpetrolera.com/showthread.php/2046-Carbonificaci%C3%B3n.
[19] http://www.metalloyd.co.uk/spanish/products_raw_materials.html
[20]http://www.oviedocorreo.es/personales/carbon/carbon_mineral/carbon%20mineral.htm
[21] http://www.pdv.com/lexico/museo/minerales/carbon.htm
[22] JOHNSON, M., HANSEN, W., RATL, W. Wave guide construction. United States Patent Ofice. Patent dec. 30, 1947.
[23] KEYSON, D. *et al.* Domestic Microwave Oven Adapted for Fast Heat Treatment of $Ba_{0.5}Sr_{0.5}(Ti_{0.8}Sn_{0.2})O_3$ Powders. Journal of Materials Processing Technology 189 (2007). Pág. 316 - 319.
[24] KLINBUN, W. RATTANADECHO, P. PAKDEE, W. Microwave heating of saturated packed bed using a rectangular waveguide (TE10 mode): Influence or particle size, sample dimension, frequency, and placement inside the guide. International Journal of Heat and Mass Transfer. Vol. 54. 2011. Pág. 1763 – 1774.
[25] LEON, Elard F. La importancia del carbón mineral en el desarrollo. Revista del Instituto de Investigaciones FIGMMG. Vol. 9, N° 18, 91 – 97. 2006. San Marcos, Perú.
[26] NAKAMURA, T., NIKAWA, Y. Precise measurement of complex permittivity of materials for telecommunications devices. Journal of Telecommunications and Information Technology. 2001.
[27] NEBOT, Eduardo. Caracterización de dieléctricos a frecuencia de microondas. Universidad de Zaragoza. 2004, 3.
[28] PICERNO, Alberto. Todo sobre hornos de microondas. Saber electrónica. Edición internacional N° 207. México, 2007. Páginas 5 – 22.
[29] PIRA, C. Densificazione di scariche a bagliore in dispositivi magnetron sputtering tramite accoppiamiento con sorgenti Hollow Cathode. Universitá degli Studi de Padova. 2007.
[30] RATTANADECHO, P. The simulation of microwave heating of wood using a rectangular wave guide: Influence of frequency and sample size. Chemical Engineering Science. Vol. 61. 2006. Pág. 4798 – 4811.
[31] RODRIGUEZ, J.M. et al. Medidas de propiedades dieléctricas de materiales de construcción utilizando una guía rectangular. Universidad Politécnica de Madrid. 2008.
[32] ROGOV, I.A. GORBATOV, A.V. Métodos físicos de procesamiento de productos

alimenticios. Industria Alimenticia. Moscú. 1974.
[33] STANLEY, R. Wave coupling arrangement. United States Patent Office. 1959
[34] STURM, G.S.J. , STEFANIDIS, G.D. , VERMEIJ, M.D. , VAN GERVEN, T.D.T. , STANKIEWICZ, A.I. Design principles of microwave applicators for small-scale process equipment. Chemical Engineering and Processing: Process Intensification. 2010. Pág 912 - 922.
[35] UNIDAD DE PLANEACION MINERO ENERGETICA. La cadena del carbón. Ministerio de Minas y Energías. República de Colombia. 2005. Bogotá, Colombia.
[36] USLU, T. ATALAY, U. AROL, A.I. Effect of Microwave Heating on Magnetic Separation of Pyrite. Colloids and Surfaces A: Physicochem. Eng. Aspects 225 (2003). Pág. 161 - 167.
[37] ZHENFENG, L. RAGHAVAN, G.S.V. ORSAT, V. Temperature and power control in microwave drying. Journal of Food Engineering 97. 2010. Pág. 478 - 483.
[38] http://www.vidrasa.com/esp/productos/duran/duran_pf.html
[39] J. B. Hasted, Liquid water: Dielectric properties, in *Water A comprehensive treatise*, Vol 1, Ed. F. Franks (Plenum Press, New York, 1972) Pág. 255-309.
[40] http://www.lsbu.ac.uk/water/microwave.html
[41] I.Takei, Dielectric relaxation of ice samples grown from vapor-phase or liquid-phase water, in *Physics and Chemistry of Ice*, ed. W. Kuhs (Royal Society of Chemistry, Cambridge, 2007) Pág. 577-584

(137c) Novel Warm Gas Multi-Contaminant Removal System

Gokhan Alptekin, Ambalavanan Jayaraman, Margarita Dubovik and Mike Cesario, TDA Research Inc., Wheat Ridge, CO

Gasification technologies convert coal and biomass feedstocks into synthesis gas feed streams that can either be used as a fuel for highly efficient power generation cycles or converted into value-added chemicals and transportation fuels. However, the coal/biomass-derived synthesis gas contains a myriad of trace contaminants (e.g. mercury, arsenic, selenium) and ammonia (NH_3) that must be removed to eliminate power plant emissions; NH_3 will convert into NO_xin the gas turbine combustor (an acid rain precursor), while Hg is designated as an hazardous air pollutant and being regulated by the U.S. EPA. These contaminants also poison the catalysts used in the manufacturing processes that converts synthesis gas into fuels or chemicals.

Although conventional clean-up technologies can effectively remove these contaminants, they typically require aqueous quenching and cooling of the syngas to around 100°F, followed by scrubbing with chemical or physical solvents, and finally absorption/adsorption of trace contaminants on solid sorbents. However, cooling the syngas and condensation of steam significantly penalizes the cycle efficiency. The power plant efficiency can be improved 3 to 4% if the synthesis gas is not cooled below its dew point because the energy in high pressure steam can be recovered by expanding it steam in the gas turbine and eliminating the energy penalty associated with the re-heating of the synthesis gas. Warm gas clean-up also eliminates the need for the expensive heat exchangers (to cool the synthesis gas to the operating temperature of the clean-up system and then to re-heat it back to before the combined cycle) and the gray water treatment problems associated with processing the large amounts of condensate.

In a DOE funded SBIR project, TDA Research, Inc. (TDA) is developing a chemical absorbent-based gas clean-up technology to remove the NH_3 and hydrogen cyanide (HCN) as well as all trace metal contaminants (e.g., Hg, As, Se and Cd) from the synthesis gas at high temperatures (500°F) in a single process step. The sorbent can be operated in a regenerable manner to remove NH_3 and Hg, while irreversibly absorbing all other contaminants. The regenerable operation for the Hg (which has very low affinity to any surface at high temperature) and NH_3 (which is present in very high concentration in the gas) ensures high utilization of the sorbent, while its high capacity to other contaminants and low cost ensures its economical use as an expendable sorbent. We will present the results from this sorbent development effort at the meeting.

(137d) "Clean" and "Dirty" Tar Reforming of Biomass Gasification Gas

Klas J. Andersson, Reforming Department, Haldor Topsøe A/S, Lyngby, Denmark and Poul Erik Højlund Nielsen, Chemicals Department, Haldor Topsøe A/S, Lyngby, Denmark

A prerequisite for the establishment of an efficient biomass-to-chemicals production or CHP applications utilizing engines is a good thermodynamic utilization of the hot synthesis gas exiting a biomass gasifier. A second and important criteria is the extraction or conversion of tars (polyaromatic hydrocarbons), which, if left untreated, leads to fouling of downstream equipment and issues around waste water treatment. Both above criteria are met by steam reforming the tar components at high temperatures in the sulfur-laden exit gas from the biomass gasifier. We discuss the benefits of different approaches depending on pressurized or near atmospheric gasification, leading to the concepts of “clean” and “dirty” tar reforming.

The basic difference between the two approaches is how dust is handled. In the dirty tar reforming processes the residual dust in the gas - after having passed cyclones - is removed downstream the tar reformer. In the clean tar reformer the dust removal takes place upstream the tar reformer and preferentially the dust removal takes place at temperatures in the vicinity of the gasifier exit temperature to minimize loss of exergy. The catalyst and processes are widely different. In the dusty tar reformer monolithic catalysts are a necessity whereas the clean reformer can use pelletized catalysts with a much higher content of active material.

We will discuss the virtues of the different approaches and look into the requirements for producing CHP, SNG or fuels like gasoline or diesel, supplemented by our achievements in clean and dirty tar reforming in industrial and pilot settings.

(138a) Synthesis of 2-Acetoxy Propanoic Acid Via Reactive Distillation an Intermediate in Synthesis of Biobased Acrylic Acid From Lactic Acid

Aspi K. Kolah, Lars Peereboom, Carl T. Lira and Dennis J. Miller, Chemical Engineering and Materials Science, Michigan State University, East Lansing, MI

The selective formation of 2-acetoxypropanoic acid (2-APA) by reaction of lactic acid with acetic acid has been carried out in continuous reactive distillation. The reaction takes place in a six-meter tall, 50 mm ID Pyrex column operated at atmospheric pressure and is catalyzed by Amberlyst-15 cation exchange resin supported in Katapak-SP11 structured packing in the middle four meters of the column. Lactic acid solutions in water were used as feedstock in the experimental study which was reacted with glacial acetic acid in the reactive distillation column. High conversions of lactic acid and its oligomers were achieved along with high selectivity to monomeric 2-APA. Experimental studies are complemented by process analysis using AspenPlus simulation software. By implementing a kinetic model and phase equilibrium data generated in our laboratories, we are able to predict the observed pilot-scale column behavior and simulate commercial-scale facilities.

The formation of 2-APA represents the first step of a process to produce acrylic acid from lactic acid. In the second step of the process, 2-APA is cracked under high temperature to form acrylic acid and liberate acetic acid that can be recycled.

Acrylic acid, its esters, and acrylate polymers are critical chemical building blocks used in a wide gamut of applications including superabsorbents, hard plastics, paints, detergents, and adhesives. The global acrylic acid market is estimated at 3.5 million tons and $8 billion annually, with a growth of 4% per year. Acrylic acid synthesized from biobased lactic acid is a drop-in replacement for the petroleum-based product; lactic acid is a mature biobased feed stock with a proven track record of high yield (>95%) from glucose and low cost.

(138b) Revolutionizing the Production of Transportation Fuels and Chemicals

David St. Angelo, Engineering, Joule Unlimited, Bedford, MA

Many technological alternatives to petroleum-based fuels have been explored, and with good reason. Limited energy supplies, pricing volatility and environmental concerns have converged to create one of the greatest challenges of our time. Joule® has an answer that surpasses biomass-derived fuels, channeling the sun's energy into a new, industry-changing category of direct solar fuels. The company's ***Helioculture™*** platform incorporates proprietary, engineered photosynthetic microorganisms to directly produce infrastructure-ready diesel, ethanol and multiple chemicals with no dependence on biomass feedstocks, agricultural land or fresh water. In parallel, Joule has developed a novel ***SolarConverter®*** system to enable the direct, continuous process with productivities greater than biomass-dependent methods, which require numerous energy-intensive steps and downstream processing to achieve an end product. In contrast, using sunlight, non-potable water and waste CO2 from industrial emitters or pipelines, Joule ultimately targets productivities of up to 15,000 gallons of diesel and 25,000 gallons of ethanol per acre annually, at stable costs as low as $50/bble and $1.28/gallon respectively (without subsidies).

(138c) Production and Upgrading Of Renewable Bio-Crude From Large Scale Algae Ponds

Neil Osterwalder, Dan Sajkowski and Ben Saydah, Sapphire Energy Inc., San Diego, CA

Renewable fuels from algae are moving closer to commercialization as Sapphire Energy showed with their Integrated Algae Bio-Refinery (IABR). The IABR in New Mexico was commissioned in 2012 and is producing algae derived bio crude oil on a continuous basis using Sapphire's proprietary conversion and extraction process. This algae bio-crude contains a wide range of hydrocarbons, fatty acids, aromatics, and nitrogen compounds and other heteroatoms. This chemical profile suggests that algal bio-crude can be economically processed by blending with existing refinery streams and co-processing the combined streams as opposed to the more costly development of algal-specific refineries. Different options of co-feeding the algae crude to a refinery can produce a full suite of refinery products from petrochemicals to gasoline, diesel and other refinery products. This study describes the products that can be produced from algae bio-crude and its implementation into existing refining infrastructure.

(138d) Membrane-Assisted Downstream Processing for Biobutanol Purification

Sebastian Heitmann, Daniel Welz, Philip Lutze and Andrzej Górak, Department of Biochemical and Chemical Engineering, Laboratory of Fluid Separations, TU Dortmund University, Dortmund, Germany

One of the main challenges for chemical industry is the change from fossil resources to a sustainable production of chemicals. The change to biobased raw materials and biotechnological processes often leads to product inhibitions of microorganisms and diluted system, resulting in an expensive downstream processing. One bulk chemical that can be produced in a sustainable way is biobutanol. One method of producing biobutanol utilizes the so-called acetone–***n***-butanol–ethanol (ABE) fermentation that can be carried out on basis of e.g. cellulosic and lignocellulosic biomass or agricultural wastes. However, the productivity of this fermentation is still limited, because ***n***-butanol is toxic towards the microorganisms used in ABE fermentation. That is why the concentration of n-butanol in the fermentation broth is often limited to values lower than 2 wt.-percent [1]. For n-butanol recovery mainly distillation processes are used which - as result of the low concentrations - have a high energy demand. About 80 % of the total energy demand of the downstream process is required for the separation of acetone, n-butanol and ethanol from the fermentation broth. Alternatively to distillation the fermentation can be coupled with a pervaporation unit, thus n-butanol can be separated from the broth continuously to prevent the n-butanol concentration from reaching the toxicity limit. Because of the mild operating conditions, pervaporation enables an energy-efficient n-butanol separation and facilitates the recycling of the broth with lower n-butanol content back to the fermenter.

Therefore, the aim of this work was to evaluate if it is more efficient to separate n-butanol by means of pervaporation or distillation or a combination of both. For this purpose extensive experimental studies were performed to characterize the permeation behavior of a commercial available poly(dimethylsiloxane) (PDMS) membrane (Sulzer Pervap^{TM} 4060). These studies provide a wide data basis for the downstream process analysis and are used to identify possible operating windows for the application of a pervaporation. An experimental comparison between the pervaporation of a binary mixture of n-butanol and water, the pervaporation of a synthetic medium and the pervaporation of real fermentation broth was carried out. The synthetic medium contained acetone, n-butanol and ethanol in mass ratios of 3:6:1 together with acetic and butyric acid, which are the solvent precursors. Additionally process parameters like feed composition, pervaporation temperature and permeate pressure were varied. The detailed experimental procedure is described elsewhere [2]. Next to the PDMS membrane, composite membrane materials containing ionic liquids are promising for separation of n-butanol from an aqueous solution. These membrane materials show advantageous permeation properties compared to conventional polymer membranes [2]. Therefore different supported ionic liquid membranes (SILMs) have been tested to evaluate their potential in n-butanol pervaporation.

Based on the experimental results a correlation has been developed describing partial fluxes of acetone, n-butanol, ethanol and water. Fluxes of acetic and butyric acid were neglected. With the help of this correlation modular-based process studies have been carried out using Aspen Custom Modeler^{TM} to find an optimal process configuration for n-butanol separation from the fermentation broth. It was found that pervaporation as stand-alone process as well as distillation is comparatively expensive. Because n-butanol should be almost completely separated from the broth, the concentration at the outlet of the membrane module becomes small, leading to a large required membrane area and a small permeate concentration of n-butanol.

In contrast to the application of distillation and pervaporation as stand-alone processes, a combination of pervaporation and distillation was found to be best suited for efficient n-butanol recovery. A possible membrane-assisted process is shown in Fig. 1. Pervaporation is used for continuous n-butanol removal in order to extend the fermentation cycles. Because the mass fraction of n-butanol $w_{B,Rec}$ in the recycle stream is only slightly smaller compared to the mass fraction of n-butanol in the fermentation $w_{B,Ferm}$, the required membrane area is comparatively small and a higher permeate concentration can be obtained. The complete removal of n-butanol from the broth is then carried out in distillation columns. In this case, the extension of the fermentation periods and the reduction of the stream to be processed in the energy-intensive distillation lead to a reduced energy demand and lowered downstream costs for n-butanol. The permeate stream and the solvent phase from the distillation can be processed further in a conventional distillation sequence. As existing biobutanol production plants mainly rely on distillation sequences, retrofit of a pervaporation thus enables an increase in the biobutanol production efficiency, while existing paid-off plants can still be used.

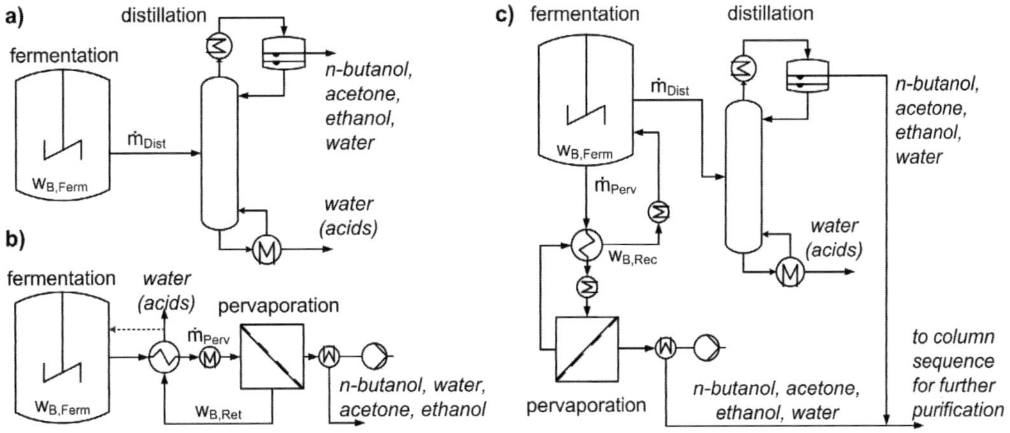

Fig. 1: Solvent removal from fermentation broth by distilliation (a), pervaporation (b) and a membrane-assisted process (c).

Acknowledgement:

The research leading to these results has received funding from the European Union Seventh Framework Programme (FP7/2007-2013) under grant agreement n° 241718 EuroBioRef. We acknowledge Sulzer Chemtech AG and the Laboratory of Biochemical Engineering, TU Dortmund University for providing PDMS membranes and fermentation broth.

[1] T.C. Ezeji, N. Qureshi, H.P. Blaschek, Bioproduction of butanol from biomass: from genes to bioreactors, Curr. Opin. Biotechnol. 3 (2007) 220-227.

[2] S. Heitmann, J. Krings, P. Kreis, A. Lennert, W.R. Pitner, A. Górak, M.M. Schulte, Recovery of n-butanol using ionic liquid-based pervaporation membranes, Sep. Purif. Technol. 97 (2012), 108-114.

Cool Planet Energy Systems and the Production of Carbon Negative Fuels

Mike Rocke, Vice President Marketing
Cool Planet Energy Systems, Inc.
mr@coolplanet.com

Abstract: Cool Planet Energy Systems is a 3-year old start-up producing gasoline blendstock from renewable non-food biomass feedstock. We have devised an economical biomass-to-liquid thermo-chemical conversion process that simultaneously produces liquid fuels and sequesterable biochar useful as a soil amendment. Fuel produced from this process is termed negative-carbon fuel since more carbon dioxide is removed from the atmosphere than released from combustion. The process for producing this fuel relies in part on a unique thermochemical decomposition of biomass that ramps the biomass temperature while the biomass is being subjected to pressure shocks. Co-produced biochar is converted into a soil amendment. We have built a 200,000 gallon per year pilot plant using this process and have demonstrated in laboratory studies that biochar generated using this process can exhibit excellent soil amendment properties. Fuel produced from this pilot plant has been fleet tested in a blend formulation in several standard configuration automobiles. The process is widely tunable in the produced biochar/fuel ratio. We also describe our plans to mass produce trailer-sized modular refinery units using this technology. We are backed by some of the world's largest energy companies and top Silicon Valley venture capitalists. We will discuss the company's trajectory from infancy to present status, and its fit in the bio-economy.

Overview

Cool Planet Energy Systems is a CA based start-up that has developed ground breaking, disruptive technology that economically converts non-food biomass into a sustainable, high-octane gasoline blend stock. Cool Planet's technology is based on twelve awarded, US patents. Hemi cellulosic or lignocellulosic biomass is processed through a mechanical biomass fractionating system that uses pressure and heat to create a series of useful volatile components that then go through proprietary catalytic systems to make high octane fuel. The process also captures the residual carbon to create a byproduct, called biochar, which is an inert, high surface area carbon that enhances soil quality through improved water and fertilizer retention and thereby increasing crop production on marginal or arid farmland. This carbon negative process removes atmospheric carbon to reverse the effects of fossil fuel consumption resulting in up to 150% carbon footprint reduction for every gallon of fuel used.

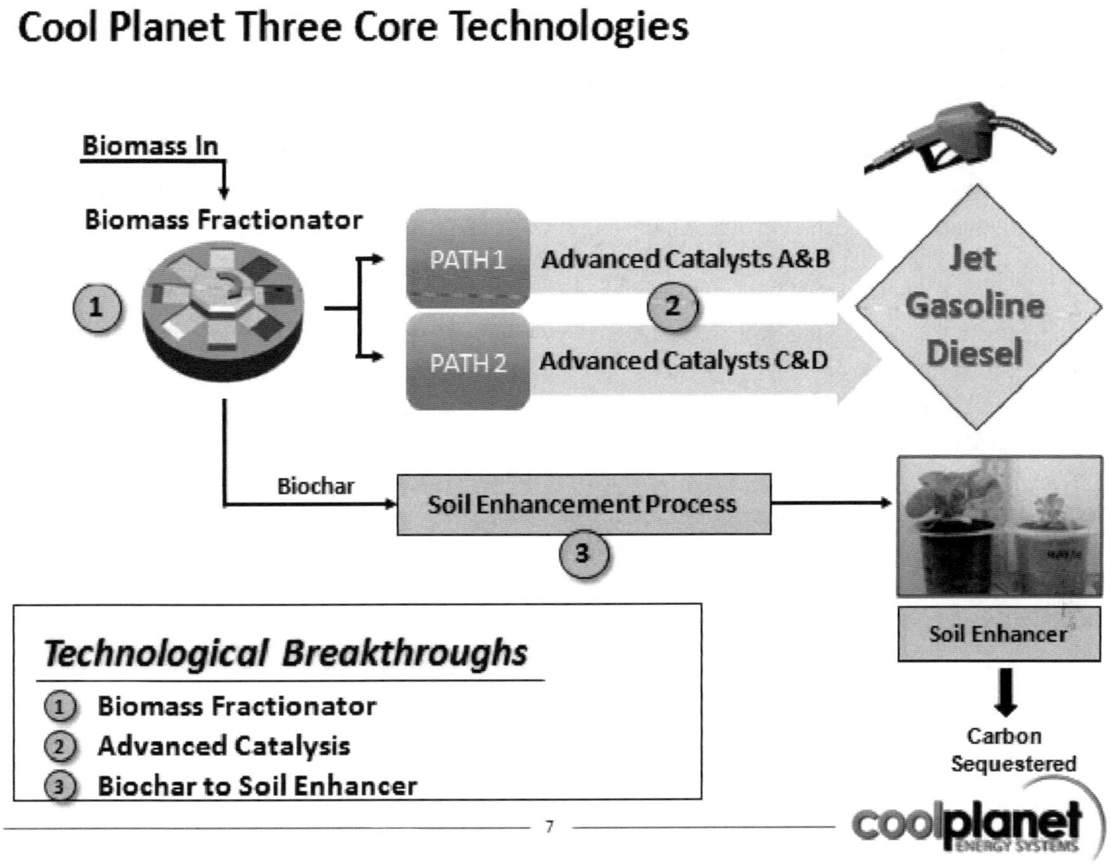

Technology of the Biomass fractionator

Ground biomass is transported by a commercial, mechanical transport system to a load hopper that pre-stages the biomass for the fractionator. Methods of introducing biomass into a processing station can include introduction via conveyor belts, hoppers, and/or pulverizers. The biomass is fed to the top of a sliding gate valve above a load station, and a precise amount of biomass is placed in a moving tray that holds the biomass for the next stage of processing. The biomass moves to the 1st heating and processing stage where the function is to drive off entrapped water, and to perform 1st step of de-hydration of the biomass while mechanically compacting it for more effective heat transfer at later stages of processing. The processing stages rotate or use a shuttle drive from step to step where the biomass is progressively heated and mechanically compacted from an initial stage temperature of 150 degrees C. As each of the following stages heat and compress the biomass, it will outgas commercially viable bio-intermediary compounds that are then taken off to catalytic reaction chambers. At each successive processing stage, the biomass is heated to a higher temperature profile and the material is compacted to a thinner layer with increasing level of elemental carbon in the remaining biomass. The processing stations thus produce volatile and non-volatile components when heated. Because the material can be heated within a relatively narrow temperature range, the non-volatile component (in the form of partially formed char) has additional volatile components embedded therein, which are extractable and processed in subsequent processing stations. The output of these processing stations is connected to an array of catalysts as described in the next section.

The final step or stage does not compact but controls the final biochar product output quantity and quality.

Advanced Catalytic Reaction Arrays

The Cool Planet system has a programmable number of processing stations, connected to an array of catalysts; a means for subjecting biomass within these stations to programmable temperatures and increments to produce a volatile and non-volatile component; a means for incrementing the non-volatile component temperature to further produce additional volatile and non-volatile components; and a means for subjecting the volatile components generated in each station through a series of catalysts. The catalysts involved are selected from the group of dehydration catalysts, aromatization catalysts, and gas-upgrading catalysts to produce different types of renewable fuel. The chemical compounds from biomass decomposition are routed to these distinct catalyst chains to produce products containing renewable, pure HC fuels and other value added products. Biomass is typically comprised of a wide array of compounds classified within the categories of cellulose, hemicelluloses, lignin, starches, and lipids. These compounds go through multiple steps of decomposition when subject to the biofractionation "pyrolysis" process. For example, hemicelluloses comprise C5 sugars such as fructose and xylose, which yield furfural and hydroxymethylfurfurals upon thermolysis. The latter compounds can be further converted to fuel intermediates furan and tetrahydrofuran. The narrow temperature windows experienced within a processing station allows for the collection of these useful intermediates.

The processing stations are connected to a series of catalyst components comprising a dehydration catalyst, an aromatization catalyst, and a gas-upgrading catalyst. The main catalysts are proprietary modified ZSM-5 catalysts. The output from each catalytic array when cooled is comprised of volatile gases, renewable fuel and water. The volatile gases are then programmed to pass through subsequent catalytic columns. Depending on the temperature (300C to 600C) of the processing stations and the biomass composition, the volatile components from any given processing station can be channeled through one or more catalysts using program selector switches.

Biochar

The last step as heat energy transfer is completed to the biomass is the production of biochar (BMF char). The BMF char remains after the previous processing steps, and the physical characteristics of the char will differ depending on the starting biomass material, which can include any of the above-identified cellulosic materials such as wood, grasses, energy crops, and agricultural waste. Different biomass feedstocks are expected to produce different types of BMF chars, varying in porosity and other physical characteristics. The biomass feedstocks can be fed individually or as mixtures of different feedstocks to produce chars containing different physical characteristics. After the BMF char is formed, a last processing step is to transfer the BMF char out of the reaction chamber for a subsequent storage or filling into a bio-char reactor for subsequent optional activation. The transfer may be accomplished via any number of mechanical means, including a press bar outfitted with a scraping knife. The Company has developed a novel process for rendering the BMF char suitable as a soil growth agent. This process comprises several steps including creating the BMF char, expelling detrimental agents within the BMF char, rendering the internal surface area of the BMF char hydrophilic, and adding suitable nutrients and microorganisms to the BMF char to promote plant growth. In addition to acting as a soil enhancing agent, the BMF char can sequester carbon for long periods of time.

Biochar Co-product – Return Carbon to the Soil

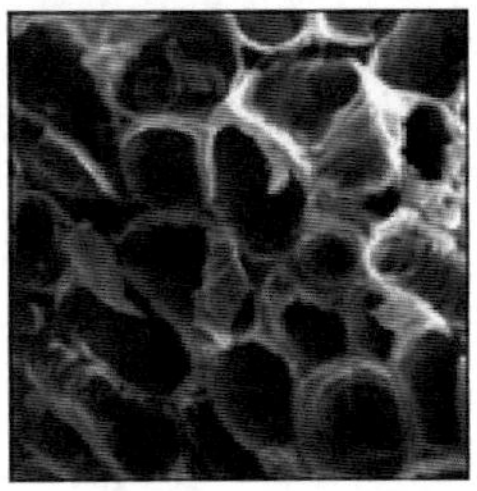

Biochar is a highly porous form of carbon produced by the selective heating of biomass, in the absence of oxygen.

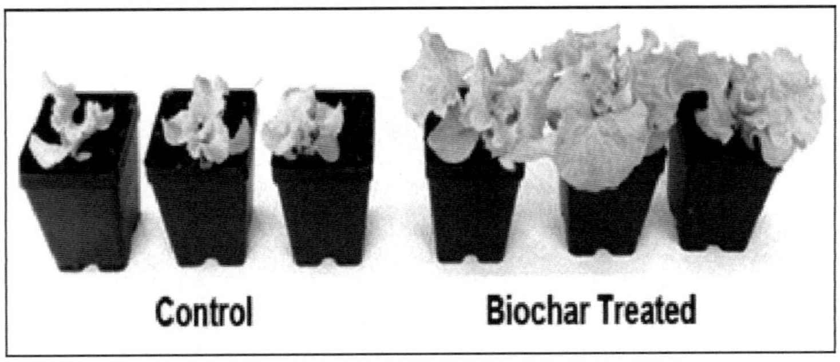

Biochar is very promising as a soil amendment that can retain water and inputs in the soil. Biochar can help maintain a nutrient rich environment and increase productivity of a wide variety of crops.

coolplanet ENERGY SYSTEMS

Drop in Hydrocarbon Fuel Results

Upon completion of catalytic conversion, the devolatilized organic components convert to various liquid fuels such as biodiesel, hydrocarbons, BTX aromatics, jet fuel, light hydrocarbons, gasoline, diesel, methanol, and DME. The organic chemicals can also be useful on their own as co-produced chemicals. By contrast, typical pyrolysis processes do not exhibit a clear volatilization profile such as this. 3rd party testing by independent laboratories shows the compounds consistent with a high octane gasoline as shown in the next graph.

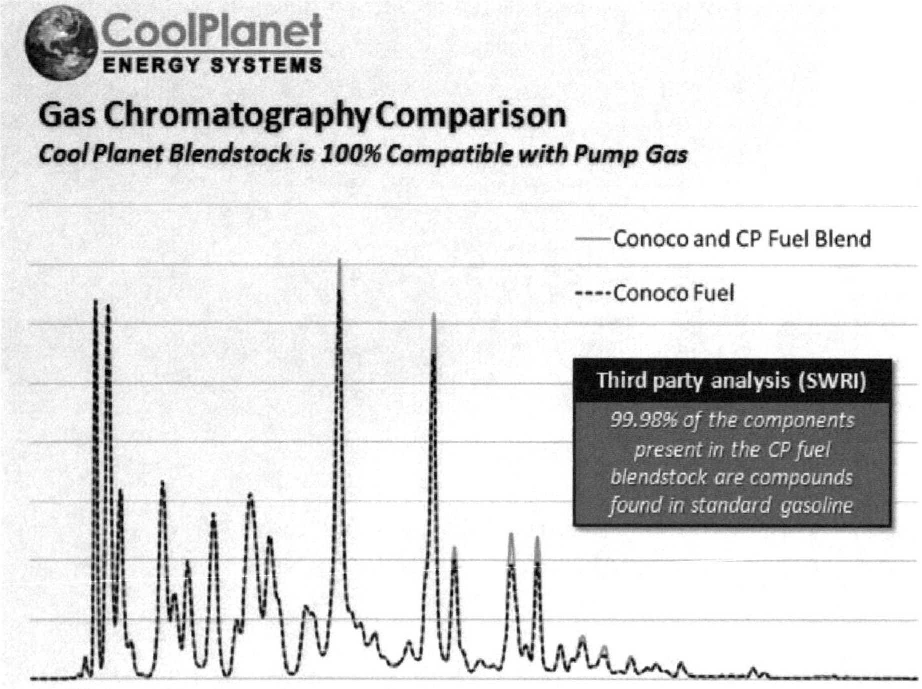

Mass Production of Distributed Units

The Cool Planet refinery equipment is modular, with fractionator and catalyst equipment connected as pairs of processing equipment. This is transported to various biomass locations where the modules are placed within a 30 mile radius of the biomass to be processed. Capex is lower by a factor of 10 compared to existing biomass refineries.

The modular refining units will be manufactured and assembled at a central plant manufacturing facility. Coordinated team will then roll out the modular facilities across the US.

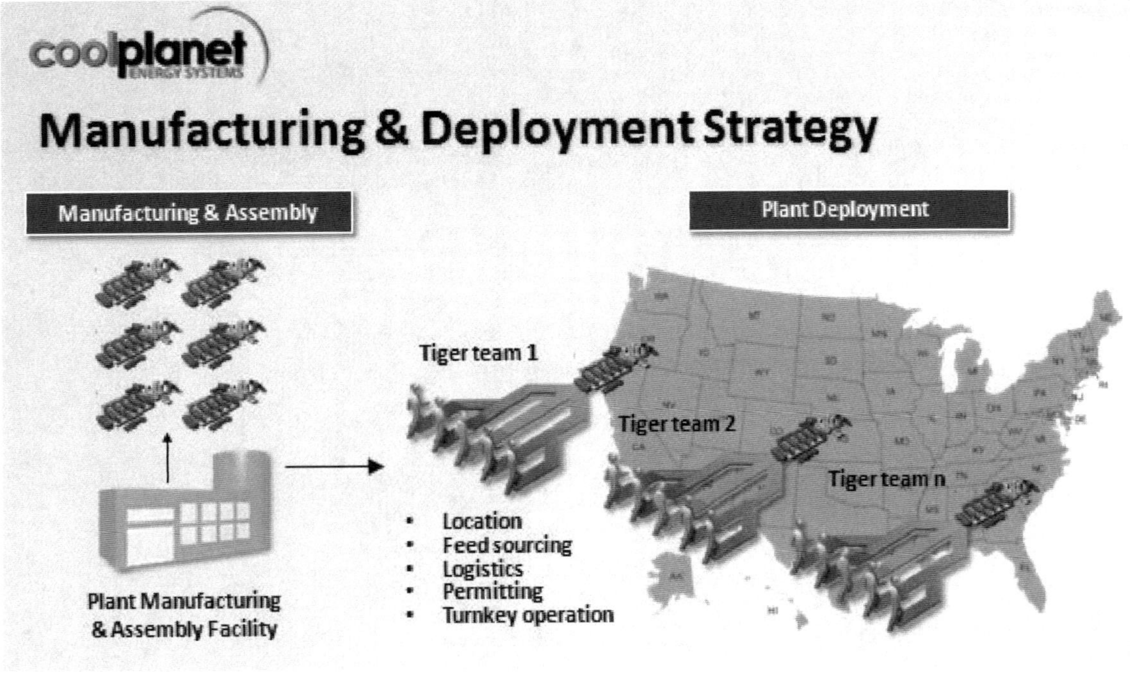

US and Global Roll-out Plan

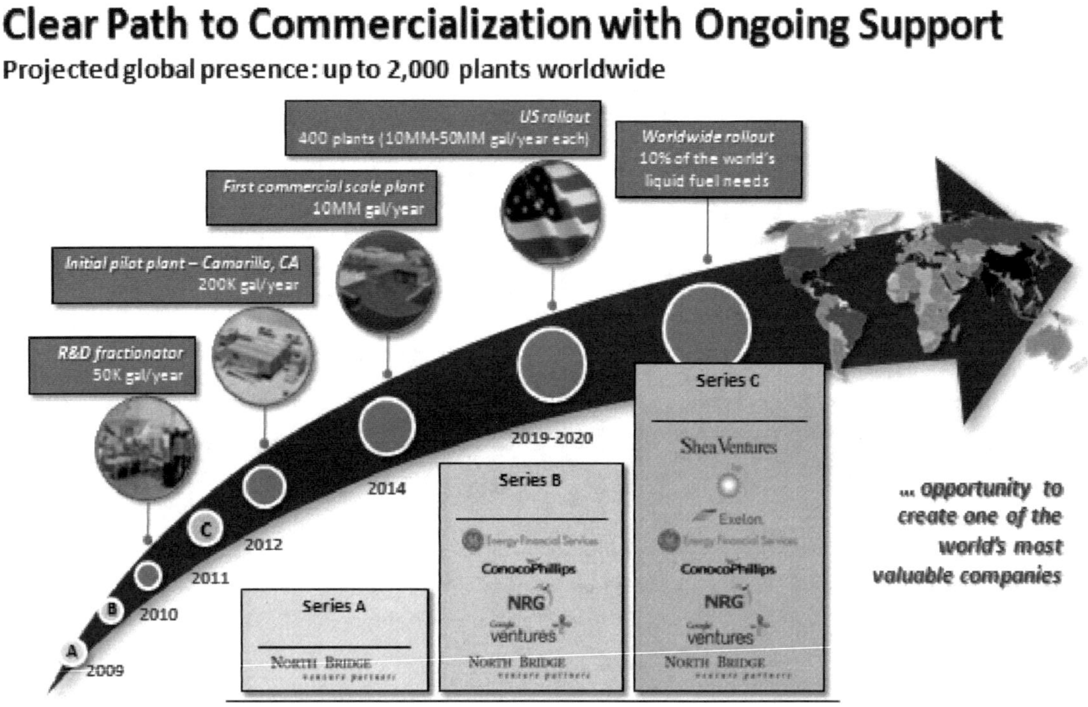

AUTHOR INDEX

AUTHOR INDEX

American Institute of Chemical Engineers
3 Park Avenue
New York, New York 10016-5991

CURRAN ASSOCIATES INC
proceedings
.com